'I really, really enjoyed reading this book. It provides a fascinating anthropological perspective on cryptozoological and related claims. The diversity of unknown animal claims considered is far wider than that seen in typical cryptozoology texts. Thus we have discussion of werebeings, non-corporeal snakes and mermaids, as well as more conventional cryptozoological fare like mystery cats.'
– Charles Paxton, University of St Andrews, UK

'When anthropologists bring their ethnographic encounters into conversation with cryptozoology the result is wildly unsettling. How are unseen creatures known? Who decides what is real? Is knowledge always tentative and provisional? What multispecies ethics do cryptids call forth? *Anthropology and Cryptozoology* is a trip through beastly domains that challenge us to think widely, generously, indeed extravagantly. '
– Deborah Bird Rose, University of New South Wales, Australia

'The scholars writing in Anthropology and Crytozoology treat the creatures of contemporary legend with a respect once reserved for ancient deities, by considering their ontological foundations and enthnozoological context. This gives new subtlety to the formerly marginal field of cryptozoology.'
– Boria Sax, author of *The Mythical Zoo* and *Imaginary Animals*

Anthropology and Cryptozoology

Cryptozoology is best understood as the study of animals which, in the eyes of Western science, are extinct, unclassified or unrecognised. In consequence, and in part because of its selective methods and lack of epistemological rigour, cryptozoology is often dismissed as a pseudo-science. However, there is a growing recognition that social science can benefit from engaging with it, for as social scientists are very well aware, 'scientific' categorisation and explanation represents just one of a myriad of systems used by humans to enable them to classify and make sense of the world around them. In many cultural contexts, myth, folk classification and lived experience challenge the 'truth' expounded by scientists. With a reflexive, anthropological approach and drawing on rich empirical and ethnographic studies from around the world, this volume engages with the theoretical and methodological issues raised by reported sightings of unrecognised animals. Bringing into sharp focus the anthropological value and challenges for methodology posed by beliefs about unclassified creatures, *Anthropology and Cryptozoology: Exploring encounters with mysterious creatures* will be of interest to anthropologists, sociologists and geographers working in the fields of research methods, anthrozoology, mythology and folklore, and human–animal interaction.

Samantha Hurn is Senior Lecturer in Anthropology and Director of the Exeter Anthrozoology and Symbiotic Ethics (EASE) working group, and the MA and Ph.D programmes in Anthrozoology at the University of Exeter, UK.

Multispecies Encounters

Series editors:
Samantha Hurn is Senior Lecturer in Anthropology and Director of the Exeter Anthrozoology and Symbiotic Ethics (EASE) working group, and Programme Director for the MA and Ph.D programmes in Anthrozoology at the University of Exeter, UK. She is the author of *Human-Animal Farm: A Multi-Sited, Multi-Species Ethnography of Rural Social Networks in a Globalised World, Humans and Other Animals: Human-Animal Interactions in Cross-Cultural Perspective* and editor of *Anthropology and Cryptozoology: Exploring Encounters with Mysterious Creatures.*

Chris Wilbert is Senior Lecturer in Tourism & Geography at the Lord Ashcroft International Business School at Anglia Ruskin University, UK. He is the co-editor of *Autonomy, Solidarity, Possibility: The Colin Ward Reader,* and *Technonatures: Environments, Technologies, Spaces and Places in the Twenty-first Century.*

Multispecies Encounters provides an interdisciplinary forum for the discussion, development and dissemination of research focused on encounters between members of different species. Re-evaluating our human relationships with other-than-human beings through an interrogation of the 'myth of human exceptionalism' which has structured (and limited) social thought for so long, the series presents work including multi-species ethnography, animal geographies and more-than-human approaches to research, in order not only better to understand the human condition, but also to situate us holistically, as human animals, within the global ecosystems we share with countless other living beings.

As such, the series expresses a commitment to the importance of giving balanced consideration to the experiences of all social actors involved in any given social interaction, with work advancing our theoretical knowledge and understanding of multi-species encounters and, where possible, exploring analytical frameworks which include ways or kinds of 'being' other than the human.

Forthcoming titles in this series

Humans, Animals and Biopolitics: The more-than-human condition
Edited by Kristin Asdal, Tone Druglitrø, Steve Hinchliffe

Animals in Place
Jacob Bull and Tora Holmberg

Anthropology and Cryptozoology

Exploring encounters with mysterious creatures

Edited by
Samantha Hurn

Routledge
Taylor & Francis Group

LONDON AND NEW YORK

First published 2017
by Routledge

2 Park Square, Milton Park, Abingdon, Oxfordshire OX14 4RN
52 Vanderbilt Avenue, New York, NY 10017

Routledge is an imprint of the Taylor & Francis Group, an informa business

First issued in paperback 2020

British Library Cataloguing in Publication Data
A catalogue record for this book is available from the British Library

Library of Congress Cataloging-in-Publication Data
Names: Hurn, Samantha, editor.
Title: Anthropology and cryptozoology / edited by Samantha Hurn.
Description: New York, NY : Routledge, 2017. | Series: Multispecies encounters
Identifiers: LCCN 2016020583 | ISBN 9781409466758 (hardback) | ISBN 9781315567297 (e-book)
Subjects: LCSH: Monsters--Psychological aspects. | Cryptozoology. | Anthropology.
Classification: LCC QL89 .A58 2017 | DDC 001.944--dc23
LC record available at https://lccn.loc.gov/2016020583

ISBN: 978-1-4094-6675-8 (hbk)
ISBN: 978-0-367-59572-2 (pbk)

Typeset in Times New Roman
by Taylor & Francis Books

Contents

Illustrations

Figure

Tables

Contributors

Luci Attala (Senior Lecturer in Anthropology, University of Wales Trinity Saint David and Ph.D. candidate, University of Exeter). Luci's research is concerned with reimagining material relationships through the lens of New Materialities. More specifically, Luci has paid attention to edibility and the coalescing themes of ingestion, incorporation, and the processes of becoming that are co-created through ingestive relationships. Currently, Luci is exploring engagements with water in rural Kenya and considers water's part in organising human bodies and social behaviours in areas where water is scarce. Luci's work in Kenya was recognised in 2014 by the United Nations with the receipt of a Gold Star Award, and in 2015 she received the Green Gown Award as Sustainability Champion in Higher Education.

Penelope (Penny) Bernard (Senior Lecturer in Anthropology, Rhodes University). Since 1997 Penny has been researching the phenomenon of the (snake/mermaid) water divinities and their role in the calling of diviner-healers. During this period she was initiated into a diviner-healer training school in KwaZulu-Natal, and this experience of radical participatory fieldwork prompted her interest in the role of dreams in guiding action, generating knowledge, and constructing notions of human–nature relatedness. Her other research interests include matters relating to health, spirit, environment, landscape, cultural heritage, kinship and morality. She has a number of publications in journals and book chapters that are the product of these research interests.

Roy Ellen (Emeritus Professor of Anthropology and Human Ecology, University of Kent). Roy's fieldwork has, over a period of forty years, focussed mainly on the Nuaulu of Seram and other peoples of the Moluccan islands of eastern Indonesia. He has also worked in Sulawesi, Java and Brunei. Throughout this period he has had a continuous professional interest in the anthropology of the environment and in ethnobiology (especially in relation to cultural cognition), and inaugurated the Kent teaching programmes in these subjects. He has also worked on the ecological and social dynamics of inter-island trading systems. Roy's current interests are focused on the

applications of cognitive anthropology to the history of science, the reproduction of Nuaulu ritual cycles, and understanding the management and
significance of cultivar diversity amongst home gardeners and farmers in
the British Isles and in the Moluccas.

Gregory Forth (Professor of Anthropology, University of Alberta). Gregory
has been conducting ethnographic fieldwork in Indonesia since 1975. His
initial and continuing interests include kinship, religion, and symbolism.
More recently his research has focused on ethnobiology, folk zoology, and
comparative ontology/epistemology. Largely on the basis of his Indonesian
research, Gregory has published several books, including *Rindi: An Ethnographic Study of a Traditional Domain in Eastern Sumba* (1981), *Dualism and
Hierarchy* (2001), *Nage Birds: Classification and Symbolism among an
Eastern Indonesian People* (2004), *Guardians of the Land in Kelimado* (2004),
and *Images of the Wildman in Southeast Asia* (2008, paperback edition
2012). In 2012, he was elected as a Fellow of the Royal Society of Canada.

Adrian Franklin (Professor of Sociology, University of Tasmania). Adrian has
been a contributor to the field of human–animal relationships since its
early days, with his book *Animals and Modern Culture* (1999), followed by
Nature and Social Theory (2002) and *Animal Nation* (2006). His work on
human–animal relations focuses on the shifts brought about by modernity;
changes in perceptions of nature, the advent of posthumanism and, in
Animal Nation, questions of nationalism, belonging and identity. His most
recent book in this area, *City Life* (2010) challenged the concept of the city as
a humanist citadel, purified of other life. His recent research has investigated
species cleansing in Australia, the mangled science of feral animals, the
benefits of companion species to human health and loneliness, and the big
cat phenomenon in the UK.

Helle V. Goldman (Chief Editor Polar Research, Norwegian Polar Research
Institute). Following fieldwork on the island of Pemba in the Zanzibar
archipelago, Tanzania, Helle received a Ph.D. in anthropology from New
York University in 1996. She then worked as a consultant for a conservation
and development project centred on Jozani Forest on Ungujaisland, Zanzibar.
This entailed, inter alia, research on the Zanzibar leopard. Now settled in
Tromsø, Norway, Helle has been the Chief Editor of the Norwegian Polar
Institute's international peer-reviewed journal *Polar Research* since 1998.
She has continued to research and write about Zanzibar and its fauna in
collaboration with a number of colleagues, as well as translating books from
Norwegian into English. Helle is currently working on a film documentary
about the Greek island where she spent much of her childhood.

Michael Heneise is a social anthropologist at the Kohima Institute, where he
is also a trustee. He has carried out fieldwork among indigenous communities
in upland South America and South Asia, most recently among the Naga in
northeast India, where he explored the relationship between dreams,

personhood, and political imagination. His interests generally encompass the intersection between ancestral knowledge, spirituality and modernity. He maintains a long-term interest in medical anthropology, the works of Maurice Merleau-Ponty, and liberation theology. He is editor-in-chief of *The South Asianist* journal, and co-editor of *The Highlander*, a new area studies journal of highland Asia. Both journals are open access, and published by the University of Edinburgh, where he received his Ph.D.

Mette M. High (Leverhulme Early Career Fellow, Department of Social Anthropology, University of St Andrews). Mette has carried out ethnographic research in Mongolia since 2001. During work for the International Labour Organization in the country, she became involved in multilateral initiatives towards improving the health and welfare of child labourers in illegal coal mines. She later began her doctoral research on the current Mongolian gold rush and received her Ph.D. in social anthropology from University of Cambridge in 2008. Her research focuses on economic transformations and moral sensibilities in extractive industries, ranging from the gold mines of Mongolia to the oil and gas fields of the USA. She has just completed her monograph entitled *Fear and Fortune: Spirit Worlds and Emerging Economies in the Mongolian Gold Rush,* and is now working on a new three-year research project entitled 'Fracking Dreams: Corporate morality and environmental politics in a new "energy economy" in the United States'.

Samantha Hurn (Senior Lecturer in Anthropology, Programme Director, Anthrozoology, and Director of the Exeter Anthrozoology and Symbiotic Ethics Working Group, University of Exeter). Sam's primary research interests fall under the umbrella category of anthrozoology (how humans think about and engage with nonhuman or other-than-human animals in a range of cultural contexts). She has conducted fieldwork in Southern Africa (Swaziland and South Africa – looking at primate conservation and human–wildlife conflict) and Europe (especially rural Andalusia, Spain and Wales, UK – focussing on domesticated animals, animals in agricultural production systems, the enrolment of animals in ritual contexts, and human kinship with dogs and other companion species). Sam has published widely on topics relating to human–animal interactions, including her book *Humans and Other Animals* (2012).

Tanya J. King (Senior Lecturer in Anthropology, Deakin University, Australia, and affiliated with the Alfred Deakin Research Institute). Tanya is an anthropologist interested in issues relating to natural resource management, the environment, gender, national identity, public policy and mythical sea creatures. She has worked with fishermen, fisheries managers and scientists, farmers, fire-fighters, water industry experts and professional rodeo riders.

Sharon Merz (Ph.D. candidate in Anthropology, University of Exeter). Sharon holds an M.A. in Death Studies from the University of Wales, Trinity Saint David. She works as a social anthropologist with SIL Togo-Bénin, a faith-based NGO that specialises in language development. Her work includes anthropological research primarily in the northwest of Benin, West Africa, where Sharon has lived since 2002. She also teaches Anthropology and advises SIL and project personnel about anthropological issues. Besides Death Studies, her interests include Anthrozoology and Cryptozoology.

Bettina E. Schmidt (Professor in the Study of Religions in the School of Theology, Religious Studies and Islamic Studies, University of Wales Trinity St David, Lampeter). She has published extensively on Caribbean and Latin American religions, identity, cultural theories and migration. Her academic interests include the anthropology of religion, diaspora identity, religious experience, urban studies, medical anthropology and gender issues. Bettina's main fieldworks were conducted in Mexico, Puerto Rico, Ecuador, New York City, and, more recently, in São Paulo, Brazil. She is the author of *Caribbean Diaspora in the USA: Diversity of Caribbean Religions in New York City* (2008, Ashgate) and co-editor of *Spirit Possession and Trance: New Interdisciplinary Perspectives* (2010, Continuum).

Stephanie S. Turner (Associate Professor of English, University of Wisconsin–Eau Claire). Stephanie teaches science communication in the Rhetoric of Science, Technology, and Culture programme. Her research interests include animal representation in visual culture, particularly cryptic and extinct fauna and taxidermy art. Her article 'Relocating "Stuffed" Animals: Photographic Remediation of Natural History Taxidermy' appeared in the spring 2013 issue of *Humanimalia*. One of Stephanie's current projects involves co-curating an exhibit of taxidermy photography and mixed media art called 'Animal Skins: Visual Surfaces' at the University of Wisconsin–Eau Claire's Foster Gallery.

Martin T. Walsh (Global Research Adviser, Oxfam). After completing a Ph.D. in 1984 based on anthropological fieldwork in southwest Tanzania, Martin lived in East Africa for many years, working as a development consultant and freelance researcher. He has been a Research Fellow in the School of African and Asian Studies in the University of Sussex, and an Affiliated Lecturer in the Department of Social Anthropology, University of Cambridge; and since 2009 the Global Research Adviser for Oxfam GB. His recent publications include 'The Not-so-great Ruaha and Hidden Histories of an Environmental Panic in Tanzania' in the *Journal of Eastern African Studies* (2012), and 'Heritage, Tourism, and Slavery at Shimoni: Narrative and Metanarrative on the East African Coast', co-authored with Stephanie Wynne-Jones, in *History in Africa* (2010).

Foreword

Conducting ethnozoological research in the 1970s on the Indonesian island of Seram, I was continually challenged by Nuaulu claims for the existence of creatures that I could not prove existed, some of which I was inclined to think were supernatural entities. These claims concerned a broad range of taxonomic categories (among them blind snakes, wild dogs, and metamorphosing lizards), as well as entities that seemed to inhabit different realms of phylogenetic space. For example, in 1970 only a handful of frog species had been scientifically reported from Seram, though there were plenty of second-hand descriptions of their appearance and behaviour in my field notes. It was only in 1975 that the herpetologist James Menzies (who had worked with Ralph Bulmer among the Kalam) and myself were able to satisfactorily match these descriptions to actual specimens, partly through the nocturnal recording of vocalisations using a directional microphone. Similarly, despite detailed Nuaulu descriptions of the Seram long-nosed bandicoot (*Rynochomelesprattorum*), it had not been scientifically observed since 1920, and its continued existence was not to be confirmed until a specimen was trapped in the 1990s on the edges of the Manusela National Park during the construction of the trans-Seram highway. My earlier field notes alone could not confirm whether creatures resembling descriptions of the bandicoot existed, whether it had once existed or even whether its status was entirely mythological. Moreover, many animals that featured in Nuaulu folk classifications did not occur locally. Yet while Nuaulu had never seen an elephant or a tiger – or until recently their photographic images – they were perfectly prepared to believe that these existed. But while none of these species likely conform to the key notion of cryptid entertained in this book, the subject-matter is directly relevant.

The catalogue of reported cryptids is impressively diverse, and given the regularities in the way they are culturally represented it is no wonder that Borges and Guerrero's *The Book of Imaginary Beings* should have become such an iconic, if playful, text. In this, and in the many historical and ethnographic accounts of cryptid morphology, a repeated and dominant theme is hybridity and anomaly. But – as anthropologists well know – these are conceptually slippery notions, since we create hybrids by recombining characters of otherwise separately distinctive species and anomalies by emphasising features that we seek to emphasise for other – often social – reasons. Indeed, anthropology has long meditated on how we reconstruct the natural world to make sense of it, personify it, how we use natural forms and the relations established between them to send moral messages, to encode relations between human groups, mark liminality and transition (as in the human-headed lions at the gates of Ninevah), and to reflect on the continuities between humans and non-humans, to get some purchase on the cosmologically remote 'other'; in essence to use such devices as a way of talking about society. And it is the ability of the imagination to exaggerate physical and behavioural differences, to make them more symbolically powerful, that is a key element in this process.

As this path-breaking book admirably demonstrates, humans inevitably perceive the world with the aid of the imagination, as well as by sharing knowledge that is not necessarily personally validated. This capacity develops from a very early age. As soon as small babies are able to discern objects within their field of sensation, they begin to associate the qualities of what we call 'animacy' with particular objects with which they interact. Commonly, the first non-human beings babies experience are not biologically animals at all but rather vocal or visual prototypes of living animals, such as 'teddy bears'. No wonder therefore that when young humans are able to match these prototypes within their emerging experience of the living world, the imagined and the 'real' begin to blur. But not only are children's imaginative worlds routinely populated with bestiaries of imagined beings, but the same core processes of cultural cognition operate in the construction of spirit forms in 'adult' cosmographies, where our experience of biodiversity provides credible templates for novel forms. As Pascal Boyer and others have argued, the same psychological apparatus used for making sense of biological reality underlies our conceptualisation of the super- and para-natural worlds.

Given the role of the imagination in constructing our memories and expectations of animal forms, and the ecological remoteness and secretiveness of many species in daily experience, it is hardly surprising that cryptozoology should prove to be such a rich field for the anthropology of science. Here is a grey area in which 'pseudo-science' assumption meets robust inferential hypotheses. Every ethnozoological account has its cryptozoological penumbrae and interstices. The puzzles these generate provide us with lessons that have a significant bearing on issues relating to the core representation and cognition of animals in general, and the relationships we establish with them. Indeed, they reveal something about the way science works as a special system of knowledge. How can the evidence before us show whether local people are making an ontological distinction between natural and supernatural, or whether they believe that something exists or not, or whether this amounts to 'belief' at all? How does 'fact' connect with 'interpretation', and where do the domains of science and ethnoscience begin and end? The history of biology is full of telling episodes that allow us to reflect on such problems. Consider, for example, that the meticulous engravings of microscopic entities in Robert Hooke's *Micrographia* (1665) were in their time no more or less credible than

Nicolaas Hartsoeker's drawing of the head of a human spermatozoon containing a tiny 'homunculus'. And compare this with Stephen J. Gould's story in *Wonderful life* (1989) of how the fossil cryptids of the Burgess shale were unravelled as being previously unknown body forms, and therefore possibly new phyla, and because of this so easily subject to fundamental misunderstandings through the available techniques of reconstruction and visual representation.

The same issues that make cryptozoology a key area for reflecting on ontology and the anthropology of science also make it important when evaluating ethnographic and other investigative methods that are otherwise taken for granted. Ethnographies of case studies of purported cryptids present a hard testing ground for how we make sense of the kind of indirect eyewitness testimony on which anthropologists extensively rely. Must we accept that such creatures exist in a scientific sense if people say they do, or if these same creatures have consequences for actions, and for the construction of other parts of a cosmography that is in other senses real and available to empirical observation? When we are suspicious of the truth claims of our informants we are inclined to dismiss their evidence as 'anecdotal', but how can people tell whether something is similar, unknown or supernatural? What does it mean to say that animals have been 'seen' or sensed in any other way when it is demonstrable that preconceptions can influence conceptions? And in penetrating such an epistemological minefield, under what circumstances might 'radical participation' ever be justified? Do we ever need more than science to validate experience?

The collection of essays brought together in this book makes a strong case for the importance of examining further the arguments outlined in this foreword, and some go beyond this, by engaging with other current connected debates, such as those pertaining to perspectival multinaturalism, animism and transspecies subjectivity. The essays demonstrate through substantive empirical cases how the critical examination of reported instances of cryptic creatures of all kinds is central to the anthropological project, and in particular to some persistent problems in ontology, epistemology and how we make sense of people's lived experience of the 'natural' world.

Roy Ellen
Emeritus Professor of Anthropology and Human Ecology,
University of Kent

Introduction

Samantha Hurn

Defining cryptozoology

Bernard Heuvelmans, the man widely dubbed the 'father of cryptozoology' (see Turner, this volume) defined cryptozoology as '[t]he scientific study of hidden animals, i.e., of still unknown animal forms about which only testimonial and circumstantial evidence is available, or material evidence considered insufficient by some!' (1988, cited in Coleman and Clark 1999: 76). Turner's opening chapter provides the historical context for the contemporary study of cryptozoology. Turner explains that the term 'cryptid' was coined in the late twentieth century to refer to the category of anomalous animals who were the purview of the emergent, hybrid (and frequently lay) discipline of cryptozoology. While cryptozoologists and their cryptid subjects existed in something of a liminal zone in academia and popular culture throughout the twentieth century, more recent discoveries of previously unknown animals such as the Komodo dragon (*Varanus komodoensis*) have reignited contemporary interest.

Along similar lines to Heuvelmans, Eberhart defined cryptozoology as 'the study of the evidence for animals that are undescribed by science' (Eberhart 2002: xlvii, cited in Walsh and Goldman, this volume). However, in their chapter Walsh and Goldman argue that science 'comprises disputed hypotheses and competing narratives' and consequently there is no definitive 'truth' to be found in the pursuit of scientific knowledge and verification alone. Surprisingly, this was a point also acknowledged by Heuvelmans himself who, despite conceiving of cryptozoology as a science, nonetheless recognised the benefits of combining scientific theories and methods with those from a disparate range of disciplines:

> [The h]idden animals with which cryptozoology is concerned, are by definition very incompletely known. To gain more credence, they have to be documented as carefully and exhaustively as possible by a search through the most diverse fields of knowledge. Cryptozoological research thus requires not only a thorough grasp of most of the zoological sciences, including, of course physical anthropology, but also a certain training in such extraneous branches of knowledge as mythology, linguistics,

archaeology and history. It will consequently be conducted more extensively in libraries, newspaper morgues, regional archives, museums, art galleries, laboratories, and zoological parks rather than in the field!

(Heuvelmans 1988)

Heuvelmans was adamant that cryptozoologists should be concerned only with real flesh and blood animals, stating that what he termed 'unexpected animals' (for example mythical creatures and ghostly apparitions) were not to be included in the discipline's remit. However, Heuvelmans' unequivocal rejection of 'unexpected animals' from the category 'cryptid', a position maintained by many prominent cryptozoologists (e.g. Coleman and Clark 1999) may have been misguided at best and ethnocentric at worst. As many of the contributions to this volume demonstrate, the reality of animals is a matter of perspective. The dividing line between zoological specimens and what Heuvelmans, Coleman, Clark and many other cryptozoologists (not to mention mainstream zoologists) would dismiss as 'paranormal' or 'Fortean' entities such as the Zanzibar leopard (*Panthera pardus adersi*) and the 'domestic' leopards kept by witches to do their bidding (Walsh and Goldman, this volume), Bengal tigers (*Panthera tigris tigris*) and *tekhumiavi* ('humans in the form of a tiger') (Heneise, this volume) or the Mongolian wolf (*Canis lupus chanco*) and *chono hün* ('wolf people') (High, this volume), is blurred or porous to say the least. The ontological turn in anthropology (see Kohn 2015 for a recent summary) has sought to raise awareness of other ways of being, of realities and worlds which may be at odds with our own (and with a 'Western', 'scientific' ontology), and which, consequently, we may struggle (or even fail) to understand.

Ontological cryptozoology

While ontology has always been at the heart of the anthropological endeavour, the 'turn' has foregrounded the many and diverse worlds inhabited not just by human groups and individuals belonging to 'other' cultures (e.g. Viveiros de Castro 1998) but also by other living beings (see, for example, Kohn 2015). Shaking the primacy of scientific knowledge based in symbolic language, the constructs of culture, and (for anthropologists in particular) a representationalist framework, an ontologically informed approach gives us cause to question Heuvelmans' definition. Indeed, while anthropologists have traditionally approached the different human groups they study as unique 'cultures' with specific belief systems which shape members' perspectives of 'the world', the ontological turn has encouraged anthropologists to see that there are, in fact, multiple worlds, none of which is more real or valid than any other. As a result, the worlds of forests (Kohn 2013), marine micro-organisms (Helmreich 2009), hyenas (Baynes-Rock 2015) and mushrooms (Tsing 2015) to name but a few, have become legitimate foci of anthropological attention. However, these non-human or other-than-human beings are more than just anthropological subjects. Indeed, these life forms think, feel and respond to their worlds in ways which also influence the lives and therefore worlds of others, including humans. Interacting or engaging with non-human others, either through the scholarly process of participant observation or through shared co-existence, enables all participants in the interaction to shape and be shaped by the interaction and the others in it – a process Ingold and Palsson (2013) have termed 'biosocial becoming'.

Cryptids, like any other non-human beings or entities, shape and are shaped by their interactions with human groups and individuals, and we are shaped by our interactions with them. The form these interactions take also has a role to play in how we come to know and understand cryptids. Through immersing themselves in the worlds where humans and cryptids co-exist or come into contact, the contributors to this volume are uniquely positioned to introduce these 'unknown' creatures. However, it is worth keeping in mind that these creatures are only 'unknown' to science and to those outside of the particular places where they are at home. For the humans (and in some cases other animals) who share those homes, these cryptids are very well known indeed.

Perspectival multinaturalism and cryptid personhood

In many cultural contexts which may, for the sake of simplicity, be classified as 'animist', it is generally recognised that 'humans and animals can move in and out of different species' perspectives by temporarily taking on alien kinds of bodies' (Willerslev 2004: 629). Moreover, according to Willerslev, in relation to the Siberian Yukaghirs in particular 'this capacity to take on the appearance and viewpoint of another species is one of the key aspects of being a person' (2004: 629). For many Amazonian peoples, interactions between animate beings are similarly governed by what Viveiros de Castro terms 'perspectival multinaturalism' (1998), a process which allows for differences between these animate beings to be transcended via what might also be termed inter-subjective interactions (Hurn 2012). For perspectival multinaturalism to work, all animate beings must recognise themselves (and others) as persons, but this personhood is, as Kohn observes (2007), dependent on the 'ontological makeup' of all concerned, which leads to creature-specific forms of personhood (jaguar personhood, vulture personhood, human personhood). Kohn gives the example of vultures for whom the smell of a rotting carcass is comparable to human appreciation of a pot of manioc tubers (2007: 7). In other words, while a rotting carcass may be disgusting for humans, for creatures with alternative ways of being in the world, such as vultures, it is deliciously appetising.

According to Ingold's earlier writings on the subject, in most 'Western' contexts, 'personhood as a state of being is not open to non-human animal kinds' (2000: 48). The so-called animal and ontological turns have demonstrated that this position is untrue for many people, and it might be suggested that the 'existence' of cryptids further challenges the primacy of human personhood in the 'Western' world also. Indeed, in many respects the human informants on whose testimonies and experiences the chapters in this volume rest – those who have encountered cryptids in various forms and situations – could be seen to be engaging in a process of perspectival multinaturalism similar to those described by anthropologists such as Viveiros de Castro (1998).

Knowing cryptids

Merz's chapter provides a clear example of this process. By recognising that the *siyawesi* bush dwarves are real in the eyes and minds (and bodies too) of her Bebelibe informants, Merz concludes they can only be real. While she may not have seen or experienced them directly herself, Merz lives with the Bebelibe, and the Bebelibe unequivocally share their world with the *siyawesi*. As one of Merz's informants concluded, 'just because science cannot prove that the *siyawesi* exist, does not mean that they do not'. This long-standing tension between science and folk knowledge was also revealed by one of Bernard's Afrikaaner informants who felt the need to come up with a scientific explanation for mermaids, combining local beliefs with knowledge of the biology of endemic, zoologically known species.

In numerous ethnographic analyses of witchcraft, sorcery, shamanism, shape-shifting, spirit possession and occult apparitions, these 'fantastical' beliefs, practices, sightings and experiences constitute social facts, but for many ethnographers they have also become lived realities. Favret-Saada and Cullen (1980), Stoller (2013) and Turner (2011) all famously underwent conversion experiences whilst researching witchcraft, sorcery and healing rituals respectively when they found their lives inexplicably and irreversibly affected by the actions of their informants or by what they experienced.

Bernard takes up this position and argues that in order to fully understand cryptids not just as our informants experience them, but as beings in their own right, anthropologists must transcend the limitations of their personal, cultural and disciplinary boundaries, and become radical empiricists. In Bernard's case this has meant consciously undergoing the rigours of training and initiation as a *sangoma*. For Attala too, experiencing the serpentine cryptids invoked during the consumption of the hallucinogenic brew *ayahuasca* necessitated that she also consumed the drug.

Admittedly not all cryptids lend themselves to a radical empiricist approach, but in his chapter Franklin nonetheless highlights the importance of empirical research when it comes to 'knowing' cryptids in general, pointing out that some existing commentaries are speculative and as a result, bark up the wrong tree when it comes to analysing sightings. For example, in his populist publication *Feral*, journalist George Monbiot (2013) suggests that big cats, along with other (what he feels are) paranormal phenomena, reflect human desires in post-modernity and asks 'Could it be that illusory big cats also answer an unmet need? As our lives have become tamer and more predictable, as the abundance and diversity of nature has declined, could these imaginary creatures have brought us something we miss?' Such a position fails to take seriously the lived experiences of the countless individuals who encounter big cats or other cryptids. For my own informants, the 'Alien Big Cats' (ABCs) they experienced were recognised as other actors within the rural landscape, despite their unexpected presence (Hurn 2009). The animals were mobilised as symbols of nationalistic resistance but more importantly, their ability to survive in an alien environment and maintain the enigma of cryptid (i.e. avoid detection, capture and the process of scientific classification) amply demonstrated their personhood and agency.

For Heneise (this volume), the piecing together of different accounts of Naga *tekhumiavi* (tiger-men) and juxtaposing them with his own subjective experiences of living with the threat of Venezuelan *nahuals* (shape shifters) in his youth, forced him to reflect on the reality of these beings who defy the laws of science. Indeed, for the anthropologists who engage with cryptids as social facts and lived realities, they, like other 'occult' beliefs, practices and beings, are recognised as important and very powerful aspects of contemporary social and political life (e.g. Comaroff and Comaroff 2003; Niehaus 2001; West 2005).

Regardless of their zoological status then, cryptids become real, become 'persons' to be known through lived experiences. Indeed, cryptids of all shapes and sizes come to life when they are encountered, and encounters take many forms, ranging from direct personal experience through to stories recounted over generations (see Hurn, this volume and Ingold 2013a; 2013b). Therefore, through knowing cryptids empirically, the contributors to this volume are in a much better place to understand them, both as significant others for human informants and as beings or entities in their own right.

Challenges to scientific knowledge and traditional ways of life

Walsh and Goldman focus their chapter on the Zanzibar leopard (*Panthera pardus adersi*), an animal classified as extinct by the international scientific community. For Zanzibaris however, the animal is still widely believed to exist both as a zoological species and as a witches' familiar. Heneise's chapter also explores the apparent contradiction between scientific and folk knowledge in Nagaland. Recent scientific research on tiger populations in northeastern India suggests that while they are Bengal tigers (*Panthera tigris tigris*), there is considerable genetic and morphological diversity between them and tiger populations elsewhere in India. However, exact population density is unknown and while tiger populations have been documented in neighbouring Assam, Arunachal Pradesh and Meghalaya, according to the National Tiger Conservation Authority(NTCA), there have been no official sightings of tigers in Nagaland prior to 2015 (Jhala et al. 2015).

Perhaps surprisingly, in some of the cases described here, the church has played an important role in keeping cryptids alive. In Merz's chapter for example, the church allows for a syncretic fusion of local beliefs in the *siyawesi* with Christian teachings. *Siyawesi* become synonymous with the devil in the local form of *Disenpode*, and in the process their role in local cosmology shifts from facilitating human civilisation at the community level (i.e. bringing agriculture to the Bebelibe) to destructive malevolence at the individual level (causing illness, death and other types of misfortune) – characteristics shared with other capricious cryptids such as *Mami Wata* (Schmidt, this volume) who intervene in the lives and fates of mortals who cross their paths and who therefore command a healthy respect or on occasion, fear. Indeed, the demonisation of liminal beings such as *siyawesi* (Merz) and mermaids (Schmidt) who are seen as dangerous (and at times specifically satanic) is also instructive, and both High's *chono* (wolf people) and Merz's *siyawesi* are divine messengers, sent to try and instil (or regain) order or to act as mediators between humans and the nonhuman realm.

Perceptions of *siyawesi* amongst individual Bebelibe as discussed by Merz are also inextricably linked to post-colonial modernity, the concomitant exposure to capitalistic ideals of materialism, individuality and, again (and most significantly), a shift away from traditional values and practices. Schmidt too notes that mermaids, and specifically *Mami Wata*, have become increasingly popular in post-colonial contexts of socio-cultural and economic flux, mass communication and increased mobility. Schmidt's chapter also reveals how cryptids can be used as political tools for creating and maintaining a sense of group identity. Indeed, for many of the contributors (e.g. Heneise, High, King) cryptids serve a significant social function in promoting (and often ensuring) adherence to traditional beliefs and practices.

Attala's chapter deals with particularly controversial cryptids: the monstrous snake apparitions who make themselves known to users of the hallucinatory brew Ayahuasca. What is particularly fascinating about this ritual process is that regardless of the background of the individual participants, visions share certain characteristics, most notably serpentine forms. Whether these snakes spew forth from the initiand's mouth or guts, or materialise from the ground beneath their feet, the serpents' presence, and the knowledge they impart, is life-changing. But more importantly, initiands experience these snakes as real, tangible entities, both during the hallucination and long after the effects of Ayahuasca have worn off. For King's shark fishers, the *ganka* is also an integral ritual actor. While they never actually 'see' the *gangka*, the creature nonetheless imparts life-changing knowledge and helps to maintain social relationships. Being on *ganka* watch initiates novice deck hands into the hierarchical but also deeply phenomenological environment of the fishing community. Anxiously searching for the *ganka* also encourages green deck-hands to become attuned to the maritime environment in ways which may not otherwise have occurred to them and which, during the long, changeable and often perilous journeys out at sea, could make the difference between life and death.

Conserving cryptids and valuing local knowledge

The controversy, in 2015, surrounding the first ever sighting by Western scientists of the 'ghost bird' or moustached kingfisher (*Actenoides bougainvillei excelsus*) by a research team from the American Museum of Natural History was, perhaps, an example of how not to engage in cryptozoological research. The scientist leading the research, Christopher Filardi, director of Pacific Programs at the Museum's Center for Biodiversity and Conservation, described the team's mission on the project blog in the language of cryptozoology, emphasising the enigmatic and ephemeral nature of cryptids:

> some species defy the familiar. There are the poorly known, reclusive animals that even when observed never fully shake the legend and mystery surrounding them. We search for them in earnest but they are seemingly beyond detection except by proxy and story. They are ghosts, until they reveal themselves in a thrilling moment of clarity and then they are gone again. Maybe for another day, maybe a year, maybe a century.
>
> (Filardi 2015: n.p.)

The urgency of finding the kingfisher was palpable in the blog and Fildari likened his team's 'discovery' of the bird to 'a creature of myth come to life' (Filardi 2015: n.p.). However, contrary to Heuvelmans' emphasis on knowing cryptids in order to protect them, Fildari and his team caught and killed their specimen. The body was deemed necessary for scientific understanding of the species, and to encourage investment in conservation. When called to account by the world's media, Filardi justified his actions on the basis of local knowledge of the bird which suggested that it was, in fact, relatively common, with a population in the region of 4,000 individuals. This indigenous information was allegedly corroborated by his team's survey, although the dead specimen was the only example of a male they saw. This particular cryptid, then, was only 'unknown' to Western scientists and not to those with whom the unfortunate 'he' co-existed.

Rather than prioritising 'science' as the arbiter of truth and the ultimate product of human social development, the accounts of the contributors to this volume reveal the currency and value of so-called 'indigenous ontologies' which do not necessarily lend themselves to scientific interpretation and analysis, but instead offer alternative ways of being in, engaging with and understanding the world. That local people know about the presence of a particular category of animal which has either been disputed by Western science or is currently unknown to us is reminiscent of colonial narratives of discovery and appropriation. Forth's chapter is a perfect example of the sort of cryptozoology which Heuvelmans had in mind when he coined the term. Using eye-witness and anecdotal testimonials, both historical and contemporary, Forth suggests in his chapter that there exists on the Indonesian island of Flores an endemic but 'mystery' felid (*ngo ngoe*) which has eluded scientific discovery but which local people recognise as being of a different order to domestic cats (*Felis catus*). Forth's chapter, along with those of several other contributors (Turner, Walsh and Goldman) emphasises the importance of indigenous or 'folk' knowledge when it comes to documenting and analysing the existence of cryptids on the ground, demonstrating how anthropologists (or ethnozoologists), in collaboration with local people, are well positioned to shed light on cryptids and interpret them in ways which utilise a combination of scientific, historical and local knowledge. Consequently the volume builds on the considerable anthropological precedent for valuing indigenous knowledge and recognising local people as ethnozoologists (or ethnoprimatologists, or ethnoethologists) (e.g. Cormier 2010; Bicker et al. 2003; cf. Sillitoe 2010). Indeed, as Hecht and Cooper (2014) observe, those humans who come into regular contact with animals can often, despite any formal training, record and comment accurately on the behavioural repertoires they observe. Not only that, but their knowledge and understanding can often rival that of academic specialists. While this certainly is the case in some contexts (e.g. Cormier 2010), Walsh and Goldman's research reveals that local people can also be unreliable witnesses, and potential conflicts of interest (e.g. illegal hunting) are difficult to mitigate against.

In his more recent work, Heuvelmans revised the remit of cryptozoology as follows; while maintaining his original emphasis on cryptids as real flesh and blood animals he argued that:

> The essential task of cryptozoology is, first, to establish a physical and behavioural identikit portrait, as precise and detailed as possible, for each apparently unknown animal about which one has significant information and then, if it is truly new, to try and discern its most probable zoological identity. Only then, knowing where, when and how to track it down, can one try with some hope of success to encounter it in nature, *in order to better know it and to protect it*.
>
> (2013: 10 emphasis added)

Heuvelmans' recognition that cryptids need protecting is also echoed by some of the contributors here, and there are numerous ethical issues at play. Bernard, for example, makes a point of not disclosing the location of her fieldwork, while King disguises the *ganka* habitat to ensure their continued existence. Consequently there emerges a conflict of interest between engaging with cryptids as 'real' and potentially exposing them to greater attention which could be damaging, as was the case for the leopards discussed by Walsh and Goodman. For Schmidt however, the representation of certain deities in cryptid form has been key to their survival. The liminality of the mermaid has facilitated the appropriation of various incarnations of water goddesses far beyond their origin in African religious traditions, enabling them to survive the journey from Africa to Brazil, North America and Europe.

In her chapter, High points out that while cryptids and nonhuman entities more generally tend to be evaluated from human perspectives, wolves are important entities in their own right. Wolves (and other animals) play significant roles in maintaining ecosystems (e.g. Monbiot 2013) while anthropogenic activities (such as mining in this case) cause significant ecological damage. The big cats encountered by Hurn and her informants in Wales have also impacted on the local ecosystem in significant ways, and their ability to act as keystone predators has been identified as a reason to protect them (2009).

So, the relationship between cryptids and conservation is complex and at times problematic. While in some instances, for example when the cryptids concerned are thought to be variants of known zoological species, the documentation of their presence is important for mobilising support and resources needed to conserve and protect habitats (e.g. Walsh and Goldman). However, the conservation agenda becomes muddied in habitats where tangible, zoologically recognised animals incarnate as ephemeral entities. In such circumstances, the conservation of the cryptid is dependent on the maintenance of local cultural traditions, and the mitigation of conflict which is often rooted in fearful coexistence. For example, in Nagaland, the first official sighting of a Bengal tiger in 2015 brought home the difficulties of living in close proximity to even small populations of such enigmatic and powerful keystone predators when a lone tigress was killed by Naga villagers in Medziphema after she entered the village and attacked livestock and a young man.

The fear of wolf people among Mongolian herders described in High's paper resonates with the fear of 'devil workers' in South America described by Taussig (2010). Denouncing individuals who behave uncharitably, or who pursue selfish agendas as *chono* ('wolf') (or as having entered into contracts with the devil in Taussig's case) serves to undermine the threat to traditional values and family units posed by capitalist enterprises and the associated commodity fetishism which can accompany the sudden influx of material wealth. A similar practice occurs in South Africa, as described in Bernard's chapter, whereby cryptids (in this case water serpents and mermaids) coerce their victims into making a pact which bestows material wealth and good fortune on individuals but with grave consequences for their kin.

While cryptids may strike fear into the hearts and minds of the humans with whom they interact, the reality is that they have much more to fear from us. In her chapter, Turner discusses a selection of creatures who embody cryptidity in different ways and uses these examples to argue that cryptids are, to use Lévi-Strauss' oft cited maxim, not just 'good to think' but also good to question in the sense that they force us to reflect on our relationships with the natural world in the Anthropocene. Because human lives impact on the lives and deaths of so many other animals, and because our actions can bring about the end of 'whole ways of life', Van Dooren (2014) suggests that narratives which seek to understand this loss might be communicated 'in a way that significantly *implicates* us – causally, perhaps emotionally, and certainly ethically' (2014: 4). That the different chapters 'tell lively stories' (Van Dooren 2014) about the lives and deaths of cryptids in a causal, emotional and ethical manner matters, because in this time of unprecedented anthropogenic activity, millions of animals die without ever being known by science.

References

Baynes-Rock, M. 2015. *Among the Bone Eaters: Encounters with Hyenas in Harar* (vol. 8). Pennsylvania: Penn State Press.

Bicker, A., Ellen, R., and Parkes, P. (eds) 2003. *Indigenous Enviromental Knowledge and Its Transformations: Critical Anthropological Perspectives.* London: Routledge.

Coleman, L. and Clark, J. 1999. *Cryptozoology A to Z. The Encyclopedia of Loch Monsters, Sasquatch, Chupacabras, and Other Authentic Mysteries of Nature.* New York: Simon and Schuster.

Comaroff, J. and Comaroff, J. L. 2003. Alien-nation: zombies, immigrants, and millennial capitalism. *The South Atlantic Quarterly*, 101(4), 779–805.

Cormier, L. A. 2010. *Kinship with Monkeys: The Guaj Foragers of Eastern Amazonia.* New York: Columbia University Press.

Eberhart, G. M. 2002. *Mysterious Creatures: A Guide to Cryptozoology.* Santa Barbara, CA: ABC-CLIO.

Favret-Saada, J. and Cullen, C. 1980. *Deadly Words: Witchcraft in the Bocage.* Cambridge: Cambridge University Press.

Filardi, C. 2015. Field journal: finding ghosts. American Museum of Natural History. 18 October 2015. (Accessed from http://www.amnh.org/explore/news-blogs/from-the-field-posts/field-journal-finding-ghosts?utm_source=social-media&utm_medium=twit ter&utm_term=20150923-wed&utm_campaign=expedition).

Hecht, J. and Cooper, C. B. 2014. Tribute to Tinbergen: public engagement in ethology. *Ethology*, 120(3), 207–214.

Helmreich, S. 2009. *Alien Ocean: Anthropological Voyages in Microbial Seas.* Berkeley: University of California Press.

Heuvelmans, B. 2013. *Natural History of Hidden Animals.* London: Routledge.

Heuvelmans, B. 1988. The sources and method of cryptozoological research. *Cryptozoology*, 7, 1–21.

Hurn, S. 2012. *Humans and Other Animals.* London: Pluto Press.

Hurn, S. 2009. Here be dragons? No, big cats! Predator symbolism in rural West Wales. *Anthropology Today*, 25(1), 6–11.

Ingold, T. 2013a. Walking with dragons: an anthropological excursion on the wild side. In *Animals as Religious Subjects: Transdisciplinary Perspectives*, edited by C. Deane-Drummond, R. Artinian-Kaiser and D. L. Clough. London: Bloomsbury, 35–58.

Ingold, T. 2013b. Dreaming of dragons: on the imagination of real life. *Journal of the Royal Anthropological Institute*, 19(4), 734–752.

Ingold, T. 2000. *The Perception of the Environment: Essays on Livelihood, Dwelling and Skill.* London: Routledge.

Ingold, T. and Palsson, G. (eds) 2013. *Biosocial Becomings: Integrating Social and Biological Anthropology.* Cambridge: Cambridge University Press.

Jhala, Y. V., Qureshi, Q. and Gopal, R. (eds) 2015. *The Status of Tigers in India 2014.* National Tiger Conservation Authority, New Delhi and The Wildlife Institute of India, Dehradun. 18 March 2016. (Accessed from http://projecttiger.nic.in/WriteRea dData/LetestNews/Document/Tiger%20Status%20booklet_XPS170115212.pdf).

Kohn, E. 2015. Anthropology of ontologies. *Annual Review of Anthropology*, 44(1), 311–327.

Kohn, E. 2013. *How Forests Think: Toward an Anthropology Beyond the Human.* Berkeley: University of California Press.

Kohn, E. 2007. How dogs dream: Amazonian natures and the politics of transspecies engagement. *American Ethnologist*, 34(1), 3–24.

Monbiot, G. 2013. *Feral: Searching for Enchantment on the Frontiers of Rewilding.* London: Penguin.

Niehaus, I. 2001. *Witchcraft, Power and Politics: Exploring the Occult in the South African Lowveld.* London: Pluto Press.

Sillitoe, P. 2010. Trust in development: some implications of knowing in indigenous knowledge. *Journal of the Royal Anthropological Institute*, 16(1), 12–30.

Singh, S. K., Mishra, S., Aspi, J., Kvist, L., Nigam, P., Pandey, P. and Goyal, S. P. 2015. Tigers of Sundarbans in India: is the population a separate conservation unit? *PloS one*, 10(4), e0118846.

Stoller, P. and Olkes, C. 2013. *In Sorcery's Shadow: A Memoir of Apprenticeship Among the Songhay of Niger.* Chicago: University of Chicago Press.

Taussig, M. T. 2010. *The Devil and Commodity Fetishism in South America.* Chapel Hill: University of North Carolina Press.

Tsing, A. L. 2015. *The Mushroom at the End of the World: On the Possibility of Life in Capitalist Ruins.* Princeton, NJ: Princeton University Press.

Turner, E. 2011. *Experiencing Ritual: A New Interpretation of African Healing.* Pennsylvania: University of Pennsylvania Press.

Van Dooren, T. 2014. *Flight Ways: Life and Loss at the Edge of Extinction.* New York: Columbia University Press.

Viveiros de Castro, E. 1998. Cosmological deixis and Amerindian perspectivism. *Journal of the Royal Anthropological Institute*, 4(3), 469–488.

West, H. G. 2005. *Kupilikula: Governance and the Invisible Realm in Mozambique.* Chicago: University of Chicago Press.

Willerslev, R. 2004. Not animal, not not-animal: hunting, imitation and empathetic knowledge among the Siberian Yukaghirs. *Journal of the Royal Anthropological Institute*, 10(3), 629–652.

1   The place of cryptids in taxonomic debates

Stephanie S. Turner

Introduction

The term 'cryptid' is a relative newcomer to the English lexicon, coined as recently as 1983, according to the *Oxford English Dictionary*, to take the place of 'sensational and often misleading terms like "monster"'. As a noun, 'cryptid' appears in popular science writing to refer to such 'improbable animals' (Museum Accepts Cryptic Collection 1999: 1079) as the coelacanth (thought to have gone extinct millions of years ago but found to exist), the Tasmanian tiger (now listed as extinct but still allegedly sighted), and the yeti (never definitively documented). Scientists writing to a scientific audience are more likely to use the adjective 'cryptic' than the noun 'cryptid', though they do so in a variety of strategically specific ways to cover much the same ground as cryptozoologists do when referring to cryptids. For example, scientists use 'cryptic' to refer to the camouflage coloration of some species, questionable hybrids, recently discovered species, undiscovered species suspected to have existed in the past, and a single species that has been found, through genetic technologies, to be multiple species.

Until recently, then, the term 'cryptid' has appeared mainly in that marginalized field mixing folklore and zoology known as cryptozoology. The so-called 'father of cryptozoology' (Coleman 2001: n.p.), Bernard Heuvelmans, used the term 'cryptid' to refer broadly to the many unknown and relict species that he was convinced still roamed the earth. A zoologist by training but quixotic by inclination, Heuvelmans focused his exhaustive research on large animals (like sea serpents) and hominoids (like the yeti), establishing a trend that has dominated cryptozoology ever since (Weidensaul 2002: 173). In this paper, the non-technical term 'hominoid' is used to refer to the taxonomic superfamily *Hominoidea*, while 'hominin' is preferred over the less precise 'hominid' to refer to those specific hominoids in the tribe *Hominini*, which includes humans and their ancestors following their split from the great apes (Stein n.d.: 4). According to Heuvelmans, cryptids matter to humans in specific ways: they have traits that are 'truly singular, unexpected, paradoxical, striking, emotionally upsetting, and thus capable of mythification' (cited in Dendle 2006: 192). It did not matter to Heuvelmans that cryptid existence lacks objective proof; circumstantial and testimonial evidence of their reality is compelling enough to take their possible existence seriously (Heuvelmans 1958: 28–29). He thus viewed cryptids as 'monsters' only in the sense that their possible existence demonstrates something of value that has been lost from the natural world because of human actions that only human representation could confirm. Since their existence depends so much on human testimony and redemption seeking, the search for cryptids takes on a moral dimension in Heuvelmans' writing. For example, at the conclusion of the study that launched the field, Heuvelmans writes,

> Tomorrow we may know one of our other relatives: the abominable snowman [yeti], for instance, who is surely a shy and gentle great ape; or perhaps an even more human primate like the tiny agogwe or the elusive orang pendek. I hope with all my heart that when he is captured there will be no needless murder. Have pity on them all, for it is we who are the real monsters.
>
> (1955: 518)

Heuvelmans' perspective continues to influence the discourse among crypto-zoologists and others on why cryptids matter. Second-generation cryptozoologist Loren Coleman's characterization of cryptids, echoing that of Heuvelmans, underscores their importance to humans: they are 'either unknown species of animals or those that are thought to be extinct but [that] may have survived into modern times and await rediscovery by scientists' (Coleman 2003: n.p.). Known only indirectly, merely suspected to exist, or somehow surviving the vicissitudes of modernity, the distinct agency of these animals challenges humans who attempt to situate them in time and place. Significantly, Coleman suggests, *any* animal lacking a precise identification or classification could be considered a cryptid (Coleman 2003). Such an expanded meaning of the term would include many more nonhumans, and not all of them megafauna: disputed type specimens, hybrids and mutants, urban wildlife, and formerly domes-ticated animals gone feral. Taking Coleman's suggestion one step further, one might also consider as cryptids such engineered microorganisms as oil-eating bacteria and synthetic genomes. Under this scheme, all sorts of anomalous nonhumans would matter that much more to humans in a variety of fields touching on human–animal studies or anthrozoology, presenting new opportu-nities to build knowledge across disciplines. Examining why and how cryptids matter illuminates the fundamental, compelling role of anomaly in both scientific and popular responses to the natural world, as well as the ways these responses inform each other. Destabilizing and disturbing, anomalies in the natural world provoke questions and drive quests. Anomalous animals function as points of negotiation over what counts as worthy knowledge in the life sciences, as their very ambiguity can help clarify the changing values and emerging problems in such projects as species cataloging and conservation efforts. Anomalous animals are also at the crux of the field of human–animal studies or anthrozoology, as our perception of their apparent monstrosity can illuminate some of the contradictions in our characterization of animal others (e.g. Fudge 2002). Putting cryptozoology into historical context and tracing its various manifestations in recent popular and professional discourse, I consider some of the ways that cryptid proliferation calls our attention to non-human agency.

Recording the cryptid moment

Why did the neologism 'cryptid' seem necessary to cryptozoologists in 1983? What was contributing to the sense that anomalous animals merited an investigative category of their own? A number of developments in establishment science were taking place at this time that lent credibility to the question of cryptid existence. One development in the early 1980s that crystallized the field was the formation of the International Society of Cryptozoology (ISC) in 1982, whose mission was 'to promote scientific inquiry, education and communication among people interested in animals of unexpected form or size, or unexpected occurrence in time or space' (Wilford 1982: n.p.). Held at the National Museum of Natural History under the auspices of the Smithsonian Institution, the society's formation and requisite journal, *Cryptozoology*, were high profile enough to command the attention of the likes of the American Philosophical Society. The attention was far from flattering, however. The eminent paleontologist George Gaylord Simpson put it bluntly: 'The pursuit of supposed mammals lacking objective evidence is not a science in an acceptable usage of that word' (1984: 1).

Yet evidently the formation of the ISC was tapping into a strong anti-establishment sentiment, as during that same year another group of academic free thinkers with equally respectable credentials organized in an attempt to shirk dogma in the sciences. The mission of this group, the Society for Scientific Exploration (SSE), was to consider 'topics which are for various reasons ignored or studied inadequately within mainstream science' (Society for Scientific Exploration 2008: n.p.), thus reflecting a perception among some scientists that the scientific enterprise could be investigating a greater variety of phenomena than it was at this time. Mentioning the formation of the ISC as a 'symptom' of late twentieth-century science's failure to investigate natural phenomena (e.g., the Loch Ness Monster) that matter to lay people, SSE member Henry H. Bauer (2002a; 2002b) railed against what he saw as the increasing hegemony of establishment science throughout the twentieth century. Science, he claimed, had moved away from the Mertonian norms of 'disinterested skepticism' in the production of 'universally valid knowledge as a public good' (Bauer 2004: 645), gradually becoming scientistic, bureaucratic, commercial, and downright fraudulent (Bauer 2004: 644). At the same time, a populist strain of distrust of establishment science was well underway as the Cold War escalated and issues like environmental pollution, food safety, and the treatment of animals in laboratories, farms, and zoos was shaping a more cautionary view of the role of science in everyday life.

Magnifying this distrust of establishment science is the awareness among cryptozoologists that 'when a knowledge domain that has potential for contributions to science is created by amateurs, it will eventually combine with and then be taken over by professionals, with the result that amateur leadership is displaced' (Regal 2009: 83. See also Dendle 2006). In turn come a loss of motivation for needed volunteers and a lack of appreciation for local knowledge. Nevertheless, a vernacular science around 'things that matter' to lay people – for example annual bird counts and county extension projects – was flourishing.

Throughout the twentieth century, dramatic cryptid discoveries and tantalizing bits of evidence had been capturing the public's attention. These ran the gamut from mythical to unexpected beasts, including a 'dragon', specifically the giant lizard known as the Komodo dragon (*Varanus komodoensis*); and a survivor from 80 million years ago, the 'living fossil' fish otherwise known as the coelacanth (*Latimeria chalumnae*) (Weinberg 2000: 28). The public's first encounters with the Komodo dragon following naturalist W. Douglas Burden's 1926 expedition to find the fabled creature must have made a considerable impression, as Burden had worked hard to craft a compelling – if not entirely naturalistic – visual narrative using carefully edited documentary film footage of his trek (Mitman 1993). Including a diorama of twelve Komodo taxidermies at the American Museum of Natural History and two live Komodo dragons on exhibit at the Bronx Zoo, all captured by Burden, Burden brought a verifiable cryptid into the public eye as never before. The public's first exposure to the coelacanth some 13 years later, though not as crafted, was equally dramatic. Instead of a 'dragon', this time a scientist had discovered what amounted to a 'living dinosaur' (Hamlin 2009a; 2009b) in the waters near the Comoros Islands, east of South Africa. Here, the compelling narrative drew on the evolutionary possibility that the ancient fish was evidence of a 'missing link' – a surviving member of a group of fishes, thought to be the ancestors of all land-dwelling creatures, 'whose fins appeared to sprout from the end of fleshy, limb-like lobes, almost like toeless legs' (Weinberg 2000: 24). Framed thus by newspapers all over the world, which were sure to include the only available photograph of the strange taxidermy, the coelacanth finding touched off an enormous public interest. With no additional sightings for more than a decade following the 1938 finding, the coelacanth tale became the kind of lost-and-found story often associated with cryptids, complete with 'wanted' posters offering a reward for anyone able to capture the alleged creature, dead or alive, and a successful expedition to find a second specimen in 1952. By 2009, there were nearly 175 coelacanth taxidermies on display around the world (Weinberg 2000: 205).

The Komodo dragon and the coelacanth are cryptids whose eventual capture, public display, and admission as legitimate subjects of study into the halls of establishment science diminished – but did not completely extinguish – their status as cryptids. The recent spate of unprovoked Komodo dragon attacks on humans, an unprecedented development in Komodo–human relations (Komodo Dragons Attacking Islanders 2009: n.p.), reifies the animal's traits as a cryptid in the sense that this new behavior is unexpected and emotionally disturbing. Similarly, the 1997 finding of a coelacanth in North Sulawesi, Indonesia, thousands of miles from the original captures in 1938 and 1952, attenuated the coelacanth's cryptid-like evasion of existing in a proper time and place. The 'cryptidity' of the Loch Ness Monster and the yeti, on the other hand, remains as unvarying as ever. In the decades leading up to the formation of the ICS, the public had become familiar with the numerous sightings of Nessie and Yeti; indeed, the serious efforts to investigate those sightings to some extent validated the field of cryptozoology (Regal 2009: 86–88). Heuvelmans' massive and lavishly illustrated 1958 book, *In the Wake of Sea-Serpents*, documented the plausibility, to many scientists of the day, of the existence of water-dwelling cryptids (particularly large, dragon-like sea creatures) familiar in folk tales. Among the French, at least, the discovery of the coelacanth solidified this conviction (Heuvelmans 1958: 26; see also Bauer 2002a: 241). Long a staple of local lore, the mysterious animal living in Scotland's largest body of water first received worldwide attention in 1933, when newswires picked up the story of a sighting reported in the *Inverness Courier*. The story triggered dozens of other people to write up their eyewitness testimonies and submit them to the Scottish papers, and later that year the first of a number of photographs of the alleged 'monster' were published. Soon the requisite investigation was launched (Weidensaul 2002: 154–155). Despite, or perhaps because of, inconclusive evidence of its existence, interest in Nessie remained strong. According to Coleman and Clark (1999: 140–142), the investigations made significant strides in the 1960s and 1970s, generating compelling – though again, inconclusive – evidence of some large, unidentifiable entity in the lake. Although Heuvelmans blamed the press for dampening scientific interest in the Loch Ness Monster (1955: 26), in fact, the media coverage triggered such a popular interest that the intrigue surrounding the cryptid, rather than the cryptid itself, became the focus (Bauer 2002b. See also Thomson 1991: 151).

Meanwhile, a similar, though more global, phenomenon was developing with regard to sightings of alleged hominoids of the 'big hairy monster' variety. The yeti, though known in various forms across Asia long before the twentieth century, became familiar in the West following the 1921 Mount Everest expedition of British explorer Charles Kenneth Howard-Bury. Howard-Bury's description of what appeared to be large, human-like footprints in the Himalayan snow ignited the public imagination, which was further aroused by a journalist's mistranslation of the Tibetan name for the creature that might have made the footprints (Sanderson 1961: 10–11; Coleman and Clark 1999: 23–24). The 'abominable snowman' thus entered the public lexicon 'like the explosion of an atom bomb' (Sanderson 1961: 11). According to cryptozoologist Ivan T. Sanderson in his history of hominoid sightings between 1860 and

1960, 'Nobody, and notably the press, could possibly pass up any such delicious term' (Sanderson 1961: 11). Indeed, in Sanderson's view, the abominable snowman account constituted 'a sort of turning point in Western thinking' (Sanderson 1961: 9) about the potential significance of 'native knowledge' (Sanderson 1961: 7), encouraging in the public consciousness a greater appreciation for the role of folk tales and the lore of amateur naturalists in adding to establishment knowledge of the natural world. At the same time, in North America, accounts of a similar creature known in Canada as 'Sasquatch' and in the United States as 'Bigfoot' began to thrive, and again the requisite searches for eyewitnesses and evidence were organized in the 1950s and 1960s. As is characteristic of the human response to cryptid manifestation, pinning a name on the unknown entity contributed to the process of its becoming a familiar yet still baffling thing, a phenomenon all the more uncanny when that thing so resembles another human being. According to Coleman and Clark, '[t]he naming of Bigfoot was a significant cultural event' (1999: 40) as it enabled mass media consumers to organize unidentifiable sights, sounds, and traces under a single rubric, a humanoid being whose possible existence had long been discussed by people in many places throughout the world. In no time, more human-like beings entered the global cryptozoological discourse – the Latin American chupacabra and the Australian yowie, among many others that have been catalogued – and while all are improbable animals, each one is characterized, in folkloric fashion, by its illumination of the distinct concerns of the people who claim to have encountered it. Indeed, according to Lauren Derby in her cultural and historical analysis of the chupacabra legend, which originated in Puerto Rico, to the Puerto Ricans, this entity represented the majority ambivalence over 'the predatory designs of the United States on Puerto Rico' (2008: 300). Regarding the yowie legend in Southern Australia, folklorist Philip A. Clarke describes its significance to various aboriginal groups as a fear response to the colonial incursion (2007: 143) while to the colonials it signified the possibility of discovering creatures previously unknown to science (2007: 147).

Locating cryptid spaces

While cryptozoologists' dedication to investigating reports of cryptids like Nessie and Bigfoot may have peaked by the mid-1970s (Regal 2009: 88), the field of cryptozoology itself had already taken root by then and indeed, the general public now seems more engrossed than ever with cryptids of all kinds. As folklorist Peter Dendle observes, '[p]opular interest in cryptozoology [...] has been fuelled by a recent publishing frenzy of encyclopaedias, dictionaries, and guides devoted to the subject, as well as by unprecedented opportunities for enthusiasts to collect data and exchange stories via the Internet' (2006: 190). In addition to its use on Loren Coleman's notable website *The Cryptozoologist*, the word 'cryptid' has crept into such popular discourse as novels, movies, and children's television, touching on the themes of suppressed knowledge, dangerous anomalies, and, most significantly, threatened nature. The novel *Cryptid: The Lost Legacy of Lewis and Clark* (Penz 2006) dramatizes the association of cryptid hominoids with political conspiracy. Bringing together a cryptozoologist, a paleontologist, and a descendant of former US president Thomas Jefferson (who sent the two explorers on their mission), the novel tells a story that, filling in the gaps in the explorers' field journals, 'threaten[s] to rewrite American history' (Penz 2006: n. p.) by exposing how the president was involved in suppressing evidence that a giant hominoid coexisted with modern humans at the time of European colonization of North America. The plot of the movie *Cryptid*, playing on the theme of the dangerous anomalous creature, tells the story of a group of scientists investigating a series of mysterious murders in South Africa attributed to a man-like creature 'previously undiscovered by science' (Cryptidthemovie. com). More empathetic to the plight of cryptids than these examples, however, is the portrayal of nature under siege in the cartoon *The Secret Saturdays*, an 'action series about the Saturdays, a family of cryptozoologists dedicated to protecting the secrets and mysteries of the world' (Cryptid Saturdays 2008–2009). Addressing such issues as climate change and biodiversity, the cartoon situates cryptids in a more altruistic context than do the forms directed at adults, thus portraying the redemptive mode of human stewardship of cryptic nature.

The current cryptid moment is occurring in spaces that cannot be dismissed as mere blogosphere banter or playground amusement. For example, a number of recent museum exhibits have explored the significance of cryptids in popular and scientific cultures. Perhaps the most significant of these was the 2006 Bates College Museum of Art exhibit 'Cryptozoology: Out of Time Place Scale', which explored the interconnectedness of lost species (like the Tasmanian tiger [*Thylacinus cynocephalus*]), found species (like the coelacanth), and imagined creatures (like Bigfoot). Describing the significance of the visual representation of animals to contemporary viewers, curator Nato Thompson identifies in their disappearance during the era of modernism and reappearance during the last 30 years 'an increasingly complicated human subject' for whom contemplation of animal others, particularly strange ones, suggests a renewed interest in examining the 'periphery of human subjectivity' (Thompson 2006: 152), that is, the possibility that what lies at the limits of our perception has a subjectivity of its own beyond our control. Like the Bates College exhibit, the traveling exhibit 'Mythic Creatures: Dragons, Unicorns and Mermaids', launched in 2007 at the American Museum of Natural History, also brought extinct animals into the same exhibit space as the legendary creatures of the exhibit's name, placing them alongside the extinct Malagasy elephant bird (*Aepyornis maximus*) and the extinct Asian primate *Gigantopithecus blacki*. Asking the obvious question, why devote natural history museum space to an exhibit focused primarily on mythic creatures?, *New York Times* critic Edward Rothstein suggests that perhaps it is 'because they allow us to glean something about how humanity struggles to make sense of the natural world'

(2007: para. 5), a task that must sometimes begin by imagining ways it might be otherwise.

The collections of cryptozoologists themselves are also on display. Bernard Heuvelmans' archive, which contains almost 1,000 books, 25,000 files, 25,000 photographs, correspondence, and artifacts, is held at the Musée de Zoologie in Lausanne, Switzerland. In 2009, the museum published the first issue of the cryptozoological journal *Kraken* in response to what it perceived to be a need for a renewed academic discussion on crypto-topics (Coleman 2009). Loren Coleman's private collection likewise includes a variety of materials, which he continues to develop via donations. Reflecting the inclusive spirit of crypto-zoology, these collections encompass every conceivable response to and representation of cryptids, including field samples, artifacts, replicas, and pop culture tchotchkes; critical reviews and commentary in academic journals; accounts of cryptids found to actually exist; and exposés of cryptid hoaxes, which cryptozoologists since Heuvelmans have been careful to document. These ways of attending to the possibility of cryptid existence as a phenomenon that is both scientific and cultural are part of the zoogeographical project to 'locate' animals in their 'proper place[s]' (Philo and Wilbert 2000: 7–8). However, given the singular nature of most cryptids, they may always remain 'out of time, place and scale' (Thompson 2006).

The sheer inclusivity of cryptozoology has a strong popular appeal in light of the increasing specialization of establishment science that developed in the late twentieth century. The periodic commentary on the phenomenon of cryptozoology that appears in mainstream scientific publications illustrates this populist appeal. Paul McCarthy's 1993 article in *The Scientist*, 'Crypto-zoologists: An Endangered Species', uses interviews with several scientists known for their cryptozoological research to show what makes cryptozoologists tick (the thrill of the quest, the lure of finding new species, an appreciation for indigenous ways of knowing) despite the professional risks involved in doing such work (lack of funding, denial of tenure, loss of credibility). Just over a decade later, however, Stephen Pincock (2004), writing in the same publication, revisits McCarthy's claim in his derisive account of 'the world's first school for cryptozoologists' started in Sweden. The expanded mission of the Global Underwater Search Team (GUST), originally founded in 1997 to look for freshwater cryptids like Nessie, now includes a six-month course in the theore-tical underpinnings of cryptozoology and a week's worth of field work. Like any good journalist, Pincock lets his interviewee, GUST founder Jan Sundberg, tell the story himself: 'We're not scientists, let me stress that, but we do try to do things in a scientific way [...]. We collaborate with the Swedish fishing board and universities' (Pincock 2004: 12). Not without irony, Pincock remarks, 'Sundberg worked for 25 years as a journalist, so he was trained to be a skeptic' (2004: 12). Like any good science magazine, *The Scientist* indulged its readers in yet another poke at cryptozoologists a year later, with Robert W. Philips (2005) claiming that cryptozoology is 'not dead yet' because of astrobiological endeavors like the Search for Extraterrestrial Intelligence

(SETI) and NASA's recent missions to look for life on Mars. According to Philips, this, too, is cryptozoological research.

Other scientific forums have taken the work of cryptozoology, and the existence of cryptids, more seriously. Mark K. Bayless' (2005) review of three cryptozoology books in the *Quarterly Review of Biology*, for example, avoids demarcating amateur and professional science, evaluating the books on the basis of the 'seriousness' of their cryptozoology and their relevance to zoologists. In the acclaimed science blog *Tetrapod Zoology*, which became an invited blog on the Science Blogs network in December 2006, and in 2008 was recognized by Networked Blogs as one of the top five zoology blogs, paleontologist Darren Naish follows news of cryptozoology and 'speculative zoology', noting that 'a substantial amount of research on the history of zoological exploration and discovery was and is cryptozoological in scope' (2007: para. 4). Like Bayless, he emphasizes that 'it's very difficult—if not impossible—to define a boundary between cryptozoology and "conventional" zoology' (2007: para. 5). However, it is in Naish's post on the continuing discoveries of mammals (Naish 2009) that his focus shifts from the philosophical problems of cryptozoology to those practical concerns over the natural world that transcend disciplinary boundaries, namely species loss, biodiversity, and conservation. His comments echo those of Robert May some 25 years earlier. In a 1984 *Nature* review of the ISC's first volume of *Cryptozoology*, May lamented the new journal's focus on hominoids and large vertebrates, not because of their exaggerated appeal, but because such focus overlooks the many smaller cryptids that are also rare and threatened: 'my reaction to *Cryptozoology* is regret … for the dissipated efforts that could be directed more productively to studying some of the species of tropical plants, insects and other organisms that may be going extinct at a faster rate than they are being classified' (May 1984: 687).

Cataloging cryptids

In light of the new knowledge of the extinctions of unidentified creatures, cryptids have, in an ironic sense befitting their ambiguity, been on the rise during the last two decades. The irony here is the post-extinction discovery and documentation of so many species that have been coexistent with humans all along. Despite all of this documented loss, the lack of knowledge of species remains considerable: estimates for the number of unknown species range from 3 million to as many as 100 million (Lövei 2001: 732; Wilson 2003: 78). What is more certain, though, is that a mass extinction of all kinds of species, both known and unknown, is currently under way. Although measuring the rate of the species loss in the current mass extinction is difficult, especially for smaller taxa like insects,

> three different methods for predicting impending extinction rates suggest future life spans of birds and mammals of 200 to 400 years if current trends

continue. These impending extinction rates are at least 10,000 times higher than background rates in the fossil record.

(Lövei 2001: 743)

This gap in our knowledge of life forms, along with the rapid rate of their extinction, indicates that an abundance of species may forever remain cryptids, known only to themselves and lost to human history and knowledge.

In contrast to estimations of the decline of biodiversity, the twenty-first century has been deemed 'a new age in biology' (Ceballos and Ehrlich 2009: 3841) because of the more than 400 new mammalian species that have been identified since 1993. Significantly, from a cryptozoological view, some of the discoveries have included 'sea monsters', and ethnographic methods are not off the table in continuing the search. For example, in response to the recent identification of several large vertebrate marine species, including a shark and three types of whale, Woodley and colleagues (2008) combine statistical regression models with an evaluation of cryptozoological and ethnozoological evidence to determine that as many as 15 more such species may yet be identified. Citing sea monsters like the merhorse and the tizhurek (a snake-like sea creature known in the Inuit oral tradition), the researchers argue that, '[b]ecause cryptozoological data are mostly discussed in the "grey literature", appraisals of these cryptids have never appeared in the mainstream literature, perpetuating a cycle whereby these putative animals remain unevaluated' (Woodley et al. 2008: 225). Though it is not as explicitly cryptozoological as the research of Woodley et al., statistical probability modelling and trend analysis (see, for example, Zapata and Robertson 2007) are common methods to estimate the number of unknown species. The use of probability methods to guess at cryptid existence – biology in a speculative mode – seems apt for species whose actuality may always be either hypothetical or confirmed only after extinction. The use of ethnographic methods may also be more necessary than some biologists would like to admit; for example, it took extensive interviewing with local Indonesian fishermen for marine biologist Mark Erdmann to verify that a fish he had spotted at market, well known to the locals as the foul-tasting but beautiful rajah laut (king of the sea), was indeed a coelacanth (Weinberg 2000: 172–178; Holder et al. 1999: 12616). It is more than a little incongruous, then, that species like the coelacanth, long known as living creatures by their human co-habitants, should be dubbed 'lazarus species' by the strangers who only knew them through the fossil record but are eager to claim naming rights of the newly 'found' species. More important than who gets credit for recognizing the return of a fossil species, however, is the action taken to protect it (see, for example, Fricke 1997 and Erdmann 2006).

Conservation efforts to protect known species are arguably another factor in the increase in cryptid sightings. Increasingly, we understand the natural world as a complex system in which the endangerment or loss of species, including plants comprising habitat, signifies some larger, typically anthropogenic, problem. To the extent that cryptid existence hinges on human activity, then, conservationist Robert May's disappointment with *Cryptozoology* was right on target; the field of conservation, from its founding in 1985 onward, has been characterized as a 'crisis discipline' requiring a prompt human response (Soulé 1985: 727; see also DeSalle and Amato 2004: 702). Michael Soulé's suggested 'rules for action' (1985: 729) to maintain natural biological systems involve a recognition of the coevolution and interdependence of species, among other axioms, while his normative postulates establish the shared values of this burgeoning field. Among them is the ecophilosophical notion that 'biotic diversity has intrinsic value irrespective of its instrumental or utilitarian value' (1985: 731). Based on Soulé's description of conservation biology, this new field was going to require interdisciplinary cooperation from sociologists, economists, and philosophers no less than taxonomists, evolutionary and molecular biologists, and ecologists. From a humanistic standpoint, crypto-zoology remains important in this interdisciplinary effort; according to Dendle, it has a direct relationship to the mass-scale anthropogenic extinction now underway: 'One important function of cryptozoology … is to repopulate liminal space with potentially undiscovered creatures that have resisted human devastation' (2006: 198). Liminal space, in Dendle's view, serves as the metaphoric meeting ground for animals past, present, and future, a conceptual coexistence that may be essential to coming up with practical solutions to the biodiversity crisis.

Soulé's call to action had a receptive audience. Heuvelmans had already made the cryptozoologists' case for biodiversity in his 1983 article 'How Many Animal Species Remain To Be Discovered?' a point that Jared Diamond reiterated in his similarly titled 1985 piece, 'How Many Unknown Species Are Yet To Be Discovered?' Acknowledging cryptozoologists as among those few people still engaged in the search for new species, Diamond's article conveys the robust spirit needed for the interdisciplinary work of cataloging and pro-tecting species. One leader in this ambitious undertaking is evolutionary biologist E. O. Wilson. Observing at the turn of the century that fewer than 2 million of the earth's potentially 10 million species have so far been identified and arguing that the 'descri[ption] and classif[ication of] all of the surviving species of the world [should] be one of the great scientific goals of the new century' (Wilson 2000: para. 2), Wilson helped launch the online *Encyclopedia of Life* (eol.org), a 'global partnership between the scientific community and the general public' intended to 'make freely available to anyone knowledge about all the world's organisms' (Wilson 2003). A searchable, constantly changing database curated by professional scientists, the project incorporates all previous species-cataloging efforts, of which there are too many to list. According to Wilson, it will enable a census of 'biotas of entire ecosystems … revealing unknown invertebrates and the smallest invertebrates, which still comprise most species yet lack even a name' (2003: 77). Wilson envisages the encyclopedia 'transform[ing] the very nature of biology' (2003: 77). On this point, his tremendous optimism casts a long shadow, for such a transformation will need to resolve a number of philosophical and technical problems that have so far been intractable: taxonomic and phylogenetic approaches and methods to describing life forms are 'simply incompatible' (Wheeler 2008: 1), even though both are needed to address the biodiversity crisis (Meier 2008: 118).

While the difficulty with bringing taxonomy and phylogeny into constructive conversation may hinder global biodiversity cataloging, it clarifies the goals of these big projects and lends credibility to cryptozoology. Taxonomy focuses on identifying type specimens through observed physical characteristics and behavior; as such, it is labor intensive and often calls upon the local knowledge of amateur or 'folk' naturalists. Taxonomic methods have traditionally been the modus operandi of cryptozoologists, who must first establish conclusive evidence for the existence of an alleged species by proffering a body. In this way, taxonomy serves as the starting point for the experimental methods of phylogeny, which seeks to articulate evolutionary processes, via lineage mapping and, especially, DNA sequencing. Applied together, taxonomic identification techniques and DNA sequencing have identified a number of cryptic species, one of the goals of biodiversity cataloging. For example, the species of North American shore bird known as the solitary sandpiper (*Tringa solitaria*) has been found, through a combined analysis of variations in morphology and DNA, to be two distinct species (Hebert et al. 2004). Similarly, a single species of skipper butterfly common in both northern and southern hemispheres has in fact turned out to be ten species (Sáez and Lozano 2005). The greater speed of molecular identification techniques over taxonomic methods – not to mention their greater prestige among biologists and their funding agencies – has been an asset to cryptozoology. Erdmann's 1997 finding of a coelacanth in Sulawesi, Indonesia, so far from the ones caught off the coast of South Africa decades earlier, raised the question among scientists as to whether a second species had been found. Identifying a number of both morphological and genetic differences between the fish in these separate locations, they were able to confirm that indeed it had (Holder et al. 1999).

Significantly, though, molecular identification techniques have been used to confirm the *non*-cryptic status of presumed cryptids, as well. Pierce et al. (2004), for example, describe the combined use of electron microscopy, biochemical analysis, and molecular techniques in determining that the 'Chilean Blob', a 13-ton octopus-like mass of tissue found washed up on a beach in Los Muermos, Chile, was in fact the partially decomposed remains of blubber from a sperm whale (*Physeter macrocephalus*). Similarly, Carr and colleagues (Carr et al. 2002) used polymerase chain reaction (PCR) to conclude that an unidentifiable carcass discovered on the coast of Newfoundland was also that of a sperm whale. What is notable in these two reports is their reference to legends of anomalous marine animals and the affiliation of those legends to scientific investigation. As Pierce et al. comment, though such legendary creatures are 'rarely' found, 'a few monsters, like the Nordic tale of the Kraken – a large and ferocious squid-like animal – may have a basis in reality', as evidenced in the 2003 discovery of the giant squid *Mesonychoteuthis hamiltoni* (Pierce et al. 2004: 125). Carr and colleagues also link catalogs of folkloric cryptids (from

Homer's *Iliad* to Elllis' [1994] *Monsters of the Sea*) to documented recent discoveries of new marine creatures, noting how the persistence of fabled cryptids and their scientific discoverability 'keep ... us alert to the possibility of "new varieties of beings" in the deeps'(Carr et al. 2002: 1). These examples show how science's systematic *de*population of liminal cryptozoological space in fact reaffirms the very possibility of cryptid occupation of that space. Moreover, some cryptids are likely to remain cryptic despite the precision techniques of DNA analysis. For example, hybrids, very old specimens, and species whose genotypes have been altered by genetic engineering or naturally occurring horizontal gene transfer challenge current molecular identification methods, such as DNA barcoding. These limitations reflect the larger challenges scientists face in reaching any consensus on the species concept itself. Considering the increase in genetic modification of organisms, new understanding of horizontal gene transfer phenomena, and some of the challenges of extracting ancient DNA from archaeological finds, the number of cryptids is not likely to diminish.

(Mis)Recognizing cryptic human kin

Of all cryptozoological projects, the liminal spaces occupied by bipedal, human-like cryptids may be the richest sites for unearthing explanations of why cryptids matter. Unlike dragons and sea creatures, these cryptids resemble us. They thus present us with the familiar, though no less uncomfortable, task of determining just how best to associate ourselves with them. For if Bigfoot and his cryptid kin are more than simply 'wildmen' [sic] and as such matter in some more-than-animal way, how then are we to position them in our family tree, where *Homo sapiens* currently occupies the topmost branch?

Recent developments in genome mapping and species identification among primates confirm just how troubling this question is. The mapping of the chimpanzee genome in 2005, for example, revealed that *Homo sapiens* and *Pan troglodytes* share 98% of DNA and almost all of the same genes (Li and Saunders 2005). With this development, a once familiar nonhuman at once became 'paradoxical, striking, [and] emotionally upsetting' to use Heuvelmans' words (cited in Dendle 2006: 192); its similarity to humans suddenly called on us to look for its differences. According to Chris Gunter and Ritu Dhand, in a *Nature* editorial accompanying the announcement of the chimpanzee genome, 'chimps are the best starting point to study not the similarities, but the minute differences that set us [humans] apart' (2005: 47) from other living things. Since then, projects to map the genomes of some of our other cousins, such as the bonobo (*Pan paniscus*) and Neanderthal (*Homo neanderthalensis*), have intensified this effort to distinguish ourselves.

At the same time, however, as science delineates known primate kin, our hominoid cryptid counterparts proliferate. In an arch letter to *Trends in Ecology and Evolution* entitled 'Molecular Cryptozoology Meets the Sasquatch', Dave Coltman and Corey Davis attempted to 'shed the hard light of modern science' (2006: 61) onto yet another series of sightings of a mysterious bipedal creature, this time in the Yukon. After amplifying a sequence of DNA obtained from a tuft of the creature's fur left at the scene of one of the sightings, the two biologists posited rather unevenly that either 'the Sasquatch might be a highly elusive ungulate that exhibits surprising morphological convergence with primates' or 'the hair [from this creature] might have originated from a real bison and be unrelated to the Sasquatch' (2006: 61). The simplest explanation being preferable, the scientists conclude that 'the identity and taxonomy of this enigmatic and elusive creature remains a mystery' (2006: 61). Whether, in this conclusion, they are referring to the 'real bison' or 'the Sasquatch' is not clear. It may be that scientists, no less than non-scientists, relish a good mystery.

One such mystery, though with a great deal more empirical evidence and a good bit of brawl, is the dispute over how to classify the 18,000-year-old remains of an apparent hominin skeleton discovered on the small Indonesian island of Flores in 2004. Called 'Flores Man' in homage to the location of its discovery, dubbed 'the hobbit' to indicate its size and suggest its familiarity to modern humans (a designation that came from the scientists themselves, rather than emerging in the popular press [Forth 2005: 16]), and given the taxonomic designation *Homo floresiensis*, the discovery compelled *Nature* writer Henry Gee to exclaim, 'cryptozoology can come in from the cold' (2004: para. 6; see also Myers 2004). Indeed, as the tiny skeletal remains piled up, so did the questions. Almost as soon as the bones had been excavated by Australian researcher Peter Brown and colleagues, who were certain they represented a new hominin species (Brown et al. 2004: 1055), they were seized by Indonesian scientist Teuku Jacob and his team, who were eager to refute their authenticity as a new species (Jacob et al. 2006: 13421). The controversy focused on skull size. According to Jacob's team, the Flores Man skull was that of a modern human; its smaller size was due, they claimed, to the insular dwarfing common among island species and, further, a possible microcephalic abnormality. The interpretation of the remains by Jacob et al. has since been at least partially discredited (Falk et al. 2007; Weston and Lister 2009), and Peter Obendorf and colleagues (2008) claim that the Flores skeletal remains are those of modern humans with their small size being attributed specifically to an endemic hypothyroidism. The debate over Flores Man's niche in the human family tree shows no sign of waning (see also Berger et al. 2008) and brain size continues to play a part.

Disagreement over Flores Man's morphology was accompanied by a dispute over the possible behavior of the being whose skull this once was. Did this cryptic miniature biped use tools? Hunt in groups? Have language? The answer to the tool question is an undeniable 'yes', as researchers found plenty of evidence of tool use at the archaeological site (Morwood et al. 2004). Again, what makes this finding so remarkable, and the other questions so compelling, is the tiny hominin's small brain (Wong 2004: para. 11). As one commentator for *Nature* put it, '[s]uch a minuscule brain in a species so recent that also made stone tools has strained credulity' (Lieberman 2009: 41). Echoing other well-known and ideological scientific debates over the link between cranial capacity and intelligence (e.g., Gould 1996; Fausto-Sterling 1992), the argument invokes cultural issues well beyond the morphological specifications and behavioral speculations in Brown's initial report in the journal *Nature* (Brown et al. 2004). Like the racism, ethnocentrism and sexism that biased brain size research throughout the twentieth century, biases of sexism and speciesism inflect the 'Flores Man' appellation and subsequent discussion about the hominin's small skull. In a move characteristic of anthropological science working in the type specimen mode, the first skeleton found was informally designated 'Flores Man' as a signifier for any possible others that might also turn up during the dig. But in fact, as Brown and his colleagues note, the skeleton's '[p]elvic anatomy strongly supports the skeleton being that of a female' (Brown et al. 2004: 1055). The elision of sex that occurred in subsequent references to the finding is not insignificant in terms of the differences that have mattered in brain research (Fausto-Sterling 1992). Moreover, the taxonomic disagreement over whether the Flores people should be classified as *Australopithecus* or *Homo* – a classification that would determine their proximity to modern humans – indicates an anthropocentric and even speciesist bias toward maintaining *Homo sapiens*' status as the only hominin.

Further ambiguating the scientific controversy over the Flores people's identification is the Indonesian folk wisdom that a group of little people by the name of ebu gogo lived on the island of Flores as recently as 200 years ago (Forth 2007: 261). As naturalist Scott Weidensaul observes, '[f]or almost every regional legend of a hairy giant, there is also at least one tradition of small, usually furry bipeds, often called "proto-pygmies" by cryptozoologists' (2002: 173–174). These smaller hominoid cryptids have been described scientifically as vestigial early humans – possibly species of *Australopithecus* – or even, more recently in human history, *Homo erectus* (Weidensaul 2002). In the case of the Flores people, the Nage villagers who live near the excavation tell of a group of tiny humans who stole their food. Calling them *ebu gogo*, meaning 'grandmother who eats everything' the Nage finally drove them away (Roberts 2004). Writing in *Anthropology Today* in 2005, Gregory remarked that cultural anthropologists have yet to appreciate the connection between local accounts of these little people and scientific interpretations of the Flores remains, yet they should, as both are stories that reflect the cultural values of the groups making claims about their meaning. 'Like hobbits', Forth explains, 'both *Homo floresiensis* and ebu gogo are products of [the] human imagination'; the challenge is to 'discover ... the true source of their resemblance' (Forth 2005: 16).

In keeping with the tendency of cryptids to remain cryptic, the true identity of *Homo floresiensis* isn't likely to be determined anytime soon. The identifying potential of DNA sequencing, as scientists have accomplished with *Homo neanderthalensis*, seems impossible since, unlike Neanderthal remains, which have been located in much colder climates than Indonesia's and thus yielded more viable genetic material, the Flores people's remains have been too badly degraded for definitive DNA sequencing. Furthermore, as indicated by the finding of a cryptic proto-pygmy in Palau, Micronesia (Berger et al. 2008), it is possible that additional such cryptids will surface, as there are many land-bridge islands in Indonesia to which hominin populations might have traveled. Then, cut off from the mainland for thousands of years after waters rose, they may have shrunk in stature (Diamond 2004). If, in fact, it can be shown that not only *Homo neanderthalensis* but also *Homo floresiensis* coexisted at some point in history, the branches near the top of the Hominidae family tree may be feeling a bit crowded.

Conclusion

To the extent that 'Flores Man' remains our poorly recognized kin, 'he' also remains a cryptid. And like so many other cryptids, Flores Man signifies that which lies beyond what taxonomic codification, genetic coding, or folkloric accounts alone can make scrutable. Indeed, the trouble with cryptids is as much epistemic as it is biological or cultural. Their almost willful insistence on occupying the liminal spaces of knowledge production remind us that knowledge itself is only ever approached, never fully attained. Cryptozoological discourse can thus serve as a space in which scientific and humanistic modes of knowing intersect, demarcating some boundaries while blurring others. Cryptid manifestations, then, at once serious and playful, awaken us to not only our responsibilities to, but also the possibilities of, our surprisingly populous world at a time when our shared coexistences are threatened.

References

Bauer, H. H. 2002a. The Case for the Loch Ness 'Monster': The Scientific Evidence. *Journal of Scientific Exploration*, 16, 225–246.

Bauer, H. H. 2002b. Common Knowledge about the Loch Ness Monster: Television, Videos, and Film. *Journal of Scientific Exploration*, 16, 455–477.

Bauer, H. H. 2004. Science in the 21st Century: Knowledge Monopolies and Research Cartels. *Journal of Scientific Exploration*, 18, 643–660.

Bayless, M. K. 2005. Review of *The Beasts That Hide From Man: Seeking the World's Last Undiscovered Animals*; *Cryptozoology: Science and Speculation*; and *Encyclopedia of Cryptozoology: A Global Guide to Hidden Animals and Their Pursuers*. *Quarterly Review of Biology*, September, 367.

Berger, L. R., Churchill, S. E., DeKlerk, B. and Quinn, R. L. 2008. Small-bodied humans from Palau, Micronesia. *PLoSOne* 3: e1780. doi:10.1371/journal.pone. 0001780. Available at: http://journals.plos.org/plosone/article?id=10.1371/journal.pone. 0001780 (accessed 8 October 2015).

Brown, P., Sutikna, T., Morwood, M. J., Soejono, R. P., Saptomo, E. W. and Due, R. A. 2004. A new small-bodied hominin from the late Pleistocene of Flores, Indonesia. *Nature*, 431(7012), 1055–1061.

Carr, S. M., Marshall, H. D., Johnstone, K. A., Pynne, L. M. and Stenson, G. B. 2002. How to Tell a Sea Monster: Molecular Discrimination of Large Marine Animals of the North Atlantic. *Biology Bulletin*, 202, 1–5.

Ceballos, G. and Ehrlich, P. R. 2009. Discoveries of New Mammal Species and Their Implications for Conservation and Ecosystem Services. *Proceedings of the National Academies of Science*, 106, 3841–3846.

Clarke, P. A. 2007. Indigenous Spirit and Ghost Folklore of 'Settled' Australia. *Folklore*, 118, 141–161.

Coleman, L. 2001. Bernard Heuvelmans (1916–2001). *The Cryptozoologist*. (Accessed 23 June 2009 from www.lorencoleman.com/bernard_heuvelmans_obituary.html).

Coleman, L. 2003. The Meaning of Cryptozoology. *The Cryptozoologist*. (Accessed 20 October 2007 from www.lorencoleman.com/cryptozoology_faq.html).

Coleman, L. 2009. New Cryptozoology Journal. *Cryptomundo*. (Accessed 16 July 2009 from www.cryptomundo.com/cryptozoo-news/new-cz-journal/).

Coleman, L. and Clark, J. 1999. *Cryptozoology A to Z: The Encyclopedia of Loch Monsters, Sasquatch, Chupacabras, and Other Authentic Mysteries of Nature*. New York: Fireside.

Coltman, D. and Davis, C. 2006. Molecular Cryptozoology Meets the Sasquatch. *Trends in Ecology and Evolution*, 21, 60–61.

Cryptid Saturdays: The Secret Saturdays News and Updates. 2008–2009. (Accessed 20 July 2009 from http://cryptidsaturdays.com/).

Dendle, P. 2006. Cryptozoology in the Medieval and Modern Worlds. *Folklore*, 117, 190–206.

Derby, L. 2008. Imperial Secrets: Vampires and Nationhood in Puerto Rico. *Past and Present*, Supplement 3, 290–312.

DeSalle, R. and Amato, G. 2004. The Expansion of Conservation Genetics. *Nature Reviews Genetics*, 5, 702–712.

Diamond, J. 1985. How Many Unknown Species Are Yet to Be Discovered? *Nature*, 315, 538–539.

Diamond, J. 2004. The Astonishing Micropygmies. *Science*, 306, 2047–2048.

Erdmann, M. 2006. Lessons Learned from the Conservation Campaign for the Indonesian Coelacanth, *Latimeria menadoensis*. *South African Journal of Science*, 102, 501–504.

Falk, D., Hildebolt, C., Smith, K., Morwood, M. J., Sutikna, T., Wayhu Saptomo, E., Imhof, H., Seidler, H. and Prior, F. 2007. Brain Shape in Human Microcephalics and *Homo Floresiensis*. *Proceedings of the National Academy of Sciences*, 104(7), 2513–2518.

Fausto-Sterling, A. 1992. *Myths of Gender: Biological Theories About Women and Men*. (2nd edn). New York: Basic Books.

Forth, G. 2005. Hominins, Hairy Hominoids and the Science of Humanity. *Anthropology Today*, 21(3), 13–17.

Forth, G. 2007. Images of the Wildman Inside and Outside Europe. *Folklore*, 118, 261–281.

Fricke, H. 1997. Living Coelacanths: Values, Eco-ethics and Human Responsibility. *Marine Ecology Progress Series*, 161, 1–15.

Fudge, E. 2002. *Animal*. London: Reaktion Books.

Gee, H. 2004. Flores, God, and Cryptozoology. *Nature News*. (Accessed 30 October 2007 from www.nature.com/news/2004/041025/full/news041025-2.html).

Gould, S. J. 1996. *The Mismeasure of Man*. New York: W. W. Norton.

Gunter, C. and Dhand, R. 2005. The Chimpanzee Genome. *Nature*, 437, 47.

Hamlin, J. F. 2009a. 'Discovery' of the Coelacanth. *Dinofish*. (Accessed 3 July 2009 from www.dinofish.com/).

Hamlin, J. F. 2009b. Famous Hollywood Character Inspired by Coelacanth. *Dinofish*. Accessed 6 July 2009 from www.dinofish.com/).

Hebert, P. D. N., Stoeckle, M. Y., Zemlak, T. S. and Francis, C. M. 2004. Identification of Birds through DNA Barcodes. *PLoS Biology*, 2, 1657–1663.

Heuvelmans, B. 1983. How Many Animal Species Remain to Be Discovered? *Cryptozoology*, 2, 1–24.

Heuvelmans, B. 1968. *In the Wake of the Sea-Serpents.* New York: Hill and Wang.

Heuvelmans, B. 1955. *On the Track of Unknown Animals.* New York: Hill and Wang.

Holder, M. T., Erdmann, M. V., Wilcox, T. P., Caldwell, R. L. and Hillis, D. M. 1999. Two Living Species of Coelacanths? *Proceedings of the National Academy of Sciences*, 96(22), 12616–12620.

Jacob, T., Indriati, E., Soejono, R. P., Hsü, K., Frayer, D. W., Eckhardt, R. B., Kuperavage, A. J., Thorne, A. and Henneberg, M. 2006. Pygmoid Australomelanesian Homo Sapiens Skeletal Remains from Liang Bua, Flores: Population Affinities and Pathological Abnormalities. *Proceedings of the National Academy of Sciences*, 103(36), 13421–13426.

Komodo Dragons Attacking Islanders. 2009. *National Geographic News.* 26 May. (Accessed 26 July 2009 from http://news.nationalgeographic.com/news/2009/05/090526-indonesia-komodo-video-ap.html).

Kraken: Une Nouvelle Revue de Cryptozoologie. 2009. Musée de Zoologie Lausanne. 14 May. (Accessed 17 July 2009 from www.zoologie.vd.ch/7_Cryptozoologie/Breves_Crypto/Kraken.html).

Li, W. H.. and Saunders, M. A. 2005. News and Views: The Chimpanzee and Us. *Nature*, 437, 50–51.

Lieberman, D. E. 2009. *Homo Floresiensis* from Head to Toe. *Nature*, 459, 41–42.

Lövei, G. L. 2001. Modern Examples of Extinction. In *Encyclopedia of Biodiversity* (vol. 2) edited by S. Asher-Levin. San Diego, CA: Academic Press.

May, R. M. 1984. Cryptozoology. *Nature*, 307, 687.

McCarthy, P. 1993. Cryptozoologists: An Endangered Species. *The Scientist*, 11 January, 12.

Meier, R. 2008. DNA Sequences in Taxonomy: Opportunities and Challenges. In *The New Taxonomy*, edited by Q. D. Wheeler. Boca Raton, FL: CRC Press, 95–127.

Mitman, G. 1993. Cinematic Nature: Hollywood Technology, Popular Culture, and the American Museum of Natural History. *Isis*, 84, 637–661.

Morwood, M. J., Soejono, R. P., Roberts, R. G., Sutikna, T., Turney, C. S., Westaway, K. E., Rink, W. J., Zhao, J. X.., van den Bergh, G. D., Due, R. A., Hobbs, D. R., Moore, M. W., Bird, M. I. and Fifield, L. K. 2004. Archaeology and Age of a New Hominin from Flores in Eastern Indonesia. *Nature*, 431, 1087–1091.

Museum Accepts Cryptic Collection. 1999. *Science*, 286, 1079.

Myers, P. Z. 2004. *Homo Floresiensis*, Flores Man. *Pharyngula*, 27 October. (Accessed 13 November 2009 from http://pharyngula.org/index/weblog/comments/homo_flor esiensis/P50/).

Naish, D. 2007. More on the Mainstreamification of Cryptozoology: Former Cryptids and Hypothetical Cryptids. *Tetrapod Zoology*, 16 October. (Accessed 29 July 2009 from http://scienceblogs.com/tetrapodzoology/2007/10/mainstreamification_of_cryp tozoology.php)

Naish, D. 2009. Over 400 New Mammal Species Named Since 1993. *Tetrapod Zoology*, 23 March. (Accessed 29 July 2009 from http://scienceblogs.com/tetrapodzool ogy/2009/03/408_post-1993_mammal_species.php).

Obendorf, P. J., Oxnard, C. E. and Kefford, B. J. 2008. Are the Small Human-Like Fossils Found on Flores Human Endemic Cretins? *Proceedings of the Royal Society*, 275, 1287–1296.

Penz, E. 2006. *Cryptid: The Lost Legacy of Lewis and Clark*. Lincoln, NE: iUniverse.

Philips, R. W. 2005. Cryptozoology: It's Not Dead Yet. *The Scientist*, 17 January, 8.

Philo, C. and Wilbert, C. 2000. Introduction. In *Animal Spaces, Beastly Places: New Geographies of Human-Animal Relations*, edited by C. Philo and C. Wilbert. New York: Routledge, 1–34.

Pierce, S. K., Massey, S. E., Curtis, N. E., Smith, G. N., Olavarría, C., and Maugel, T. K. 2004. Microscopic, Biochemical, and Molecular Characteristics of the Chilean Blob and a Comparison with the Remains of Other Sea Monsters: Nothing but Whales. *The Biological Bulletin*, 206(3), 125–133.

Pincock, S. 2004. Tales from the Cryptozoologists. *The Scientist*, 8 November, 12.

Regal, B. 2009. Entering Dubious Realms: Grover Krantz, Science, and Sasquatch. *Annals of Science*, 66, 83–102.

Roberts, R. 2004. Villagers Speak of the Small, Hairy Ebu Gogo. *The Telegraph*, 28 October. (Accessed 11 November 2009 from www.telegraph.co.uk/news/world news/asia/indonesia/1475280/Villagers-speak-of-the-small-hairy-Ebu-Gogo.html).

Rothstein, E. 2007. Exploring the Nature of the Unnatural. *New York Times*, 25 May. (Accessed 20 July 2009 from www.nytimes.com/2007/05/25/arts/design/25myth.html?_r=3&scp=10&sq=chupacabra&st=cse).

Sáez, A. G. and Lozano, E. 2005. Body Doubles. *Nature*, 433, 111.

Sanderson, I. T. 1961. *Abominable Snowmen: Legend Come to Life*. Philadelphia, PA: Chilton.

Simpson, G. G. 1984. Mammals and Cryptozoology. *Proceedings of the American Philosophical Society*, 128, 1–19.

Society for Scientific Exploration. 2008. SSE Mission Statement. (Accessed 30 June 2009 from www.scientificexploration.org/mission.html).

Soulé, M. E. 1985. What is *Conserv Biol*? *Bioscience*, 35, 727–734.

Stein, P. L. n.d. Hominin or Hominid? What's in a Name? n.d. (Accessed 17 July 2009 from www.anthro.utah.edu/PDFs/courses/broughton/stein.pdf).

Thompson, N. 2006. *The Call of the Wild. Cryptozoology: Out of Time, Place, Scale*. Lewiston, ME/Zurich, Switzerland: JRP/Ringier.

Thomson, K. S. 1991. *Living Fossil: The Story of the Coelacanth*. New York: Norton.

Weidensaul, S. 2002. *The Ghost with Trembling Wings: Science: Wishful Thinking, and the Search for Lost Species*. New York: North Point Press.

Weinberg, S. 2000. *A Fish Caught in Time: The Search for the Coelacanth*. New York: HarperCollins.

Weston, E. M. and Lister, A. M. 2009. Insular Dwarfism in Hippos and a Model for Brain Size Reduction in *Homo Floresiensis*. *Nature*, 459, 85–88.

Wheeler, Q. D. 2008. Introductory: Toward the New Taxonomy. In *The New Taxonomy*, edited by Q. D. Wheeler. Boca Raton, FL: CRC Press, 1–17.

Wilford, J. N. 1982. Society Formed to Bring Them Back Alive. *New York Times*, 19 January. (Accessed 2 July 2009 from www.nytimes.com/1982/01/19/science/society-formed-to-bring-them-back-alive.html?sec=health&&n=Top%2FNews%2FScience%2FTopics%2FAnimals).

Wilson, E. O. 2003. The Encyclopedia of Life. *Trends in Ecology and Evolution*, 18, 77–80.

Wilson, E. O. 2000. A Global Biodiversity Map. *Science*, 289, 2279.

Wong, K. 2004. Digging Deeper: Q and A with Peter Brown. *Scientific American*, 27 October. (Accessed 27 October 2007 from www.sciam.com/article.cfm?articleID=00082F87-7D35-117E-BD3583414B7F0000&pageNumber=3&catID=4).

Woodley, M. A., Naish, D. and Shanahan, H. P. 2008. How Many Extant Pinniped Species Remain to Be Described? *Historical Biology: An International Journal of Paleobiology*, 20, 225–235.

Zapata, F. A. and Robertson, D. R. 2007. How Many Species of Shore Fishes Are There in the Tropical Eastern Pacific? *Journal of Biogeography*, 34, 38–51.

2 Cryptids, classification and categories of cats

An ethnozoological study of unidentified felids from eastern Indonesia

Gregory Forth

Introduction

Cryptozoological species, or 'cryptids', are categories of animals which some group of humans considers real but are not attested, or have yet to be attested, by international zoology. Occasionally, anthropologists and ethnobiologists (or, more specifically, ethnozoologists) record local representations of animals, or what sound like animals, that do not immediately match known scientific species. Ethnozoologists might subsequently find a match, but they might not. In the latter case, anthropologists, to mention only these, regularly conclude they have recorded a name for a spirit or a fantastic being.

Sometimes, to a western ear, such unidentified categories do indeed sound fantastic. The 'Loch Ness monster' is one example. The Okapi (*Okapia johnstoni*), before it was 'discovered' by Europeans early in the twentieth century, might be another. However, these are not the only kind of cryptid an ethnozoologist might learn about. Another sort would be an animal of a known type which is, so to speak, not supposed to be where people say it is. Pumas (or cougars, *Puma concolor*) in the eastern United States (Downing 1984) or similar big cats in Wales (Hurn 2009 and this volume) or the southwest of England (Franklin, this volume) would be instances of this kind of cryptid. Often, though, the possible referent of a locally recognised cryptid is not immediately obvious: a category, especially if it does not sound supernatural or zoologically implausible (as would a creature reputedly capable of walking through walls, changing its form, or speaking a human language), could be either a new species, a known species not previously documented in the locality, a documented species which has somehow been misrepresented or represented in a way that makes it sound like something very different – or, indeed, it may be completely fictitious.

Explaining the emergence and persistence of categories of the last two sorts might be thought a special preserve of anthropology. Yet anthropologists have been neither particularly enthusiastic about nor adept at developing explanations of categories whose ontological status in international science is uncertain or which otherwise possess a degree of zoological plausibility. Of course, formulating a sociological or symbolic explanation for a cultural representation – the usual anthropological recourse – does not in itself prove that what is represented has no basis in empirical reality (see Forth 2004: 115–137). By the same token, sociological or symbolic interpretations of manifestly or possibly non-empirical attributes can be equally valid when the phenomenal referent is an unequivocally known animal (for example, the pangolin among the Lele, lemurs identified as spirits of the dead among Malagasy peoples, and snakes, considered as embodiments of earth spirits in many parts of the world [see Attala, this volume]). Nevertheless, social scientists tend generally to assume that such interpretations discredit or at least reduce the zoological credibility of culturally represented images. The fact of ontological ambiguity – not knowing in advance whether the referent of a local category is empirically real or not real – may in itself be sufficient to dissuade anthropologists from becoming involved in cryptozoological questions (and it would not be difficult to argue that, for a number of reasons, they are well advised to continue this avoidance!). Yet in view of the privileged access anthropologists have to local communities through long periods of co-residence and participation of various kinds, they should be well placed to shed special light on cryptozoological problems. Ethnozoologists, who are often simultaneously social or cultural anthropologists, should be able to provide even greater insight, given their access to local environments and local zoological knowledge.

The focus of this chapter is the way the Nage people, who reside in the central part of the eastern Indonesian island of Flores, classify cats. More particularly, I am concerned with the identity of a large felid that appears not to fit any Flores species currently recognised by international zoologists. A major aim is to demonstrate the value non-western zoological knowledge, or 'folk zoology', may hold not simply for cryptozoology (which surely can never be an end in itself) but indeed for international zoology. As I show especially with reference to local categories of felids, Nage representations of what, to a western eye, may appear as 'cryptids' contradict a view, implicit in a good deal of modern anthropology – but, ironically, sharing much in common with older European representations of 'primitives' – that members of small-scale, non-western societies are governed by an unbridled imagination and so typically formulate and express their thought concerning non-human animals, and even organise their behaviour in relation to these, on the basis of all sorts of fantastic notions which are either empirically false or have little grounding in reality.

Hairy hominoids and far-ranging dragons

During ethnographic fieldwork conducted over the last thirty years on Flores, I have come across two other sorts of crypo-categories besides the mystery cats. Such categories might be called 'ethno-cryptids', but only insofar as local folk zoologists themselves are unsure of their status as empirical creatures or how they might be related to known ethnotaxa. One sort of crypto-category – the

Loch Ness monster variety as it were – concerns what would appear to comprise non-sapiens hominoids or hominins. In some parts of the island, these creatures are described as currently extinct, and one category reckoned to be extinct (a hominoid named *ebu gogo*; see Forth 2012a: 12–49; Forth 2012b; and Turner, this volume) has been hypothetically identified with the recently discovered sub-fossil hominin interpreted as the type specimen of a new species, *Homo floresiensis*. Images of this sort, from Flores and other parts of Southeast Asia, are the subject of a recently published book (Forth 2012a). What light anthropology or ethnozoology might shed on such 'wildmen', as I have previously chosen to call them, is not my major concern in what follows – largely because the matter has been discussed elsewhere. Indeed, for the moment I shall say no more about them, though they will briefly come up again in my conclusions.

The other kind of 'cryptid' reported by people from several parts of Flores exemplifies a known species seemingly out of place – or, rather, did so until its existence in previously unsuspected places was established by international scientists. This is the famous Komodo monitor, or 'Komodo dragon' (*Varanus komodoensis*). The dragons are interesting because, prior to 1912, they were known empirically only to the human inhabitants of Komodo Island and neighbouring islands in eastern Indonesia. Their scientific discovery followed a report by a Dutch administrator, van Steyn van Hensbroek, who had received information from Flores Islanders that the lizards occurred not only on Komodo but also in the vicinity of Labuan Bajo, in the far western part of Flores (Ouwens 1912: 1). Until 1980, the western Flores region of Manggarai, in which Labuan Bajo resides, was regarded as the eastern extent of the Komodo dragon's range. However, from the 1910s and 1920s, ethnographers – or, more exactly, amateur ethnographers in the shape of Dutch colonial officials and missionaries – began recording local accounts of the huge lizards from parts of Flores much further to the east (see Van Suchtelen 1921; De Jong 1937).

When I myself began research on Flores in 1984, local people claimed that Komodo dragons, although rare, could be found in various parts of north central and northeastern Flores. The easternmost location was an area just west of the eastern Flores port town of Maumere, the island's largest settlement. That Komodo dragons could occur this far away from Komodo Island seemed unlikely, even somewhat fantastic. It also suggested possible confusion with another monitor species, the much smaller water monitor (*Varanus salvator*), a common animal found throughout Flores and on most other eastern Indonesian islands. However, in 1985 a team of Indonesian researchers did indeed discover Komodo dragons in northeastern Flores, in a region not far to the west of the port town of Maumere, where they managed to capture four specimens (Ciofi and Gibson 2006). This was in the vicinity of Ndondo, from where H. R. Rookmaker, a Dutch colonial administrator, recorded local reports of Komodo dragons being captured, apparently by natives, in 1926 (De Jong 1937).

Adding to these older reports, accounts I recorded from villagers in 2008 provide strong circumstantial evidence for a population of Komodo dragons inhabiting the Golo Nio region of north central Flores (Forth 2010). Golo Nio is one place where international zoologists have yet to conduct research or obtain specimens of *Varanus komodoensis*; hence attention to the claims of Golo Nio people may well repay such a research effort in the future, especially as the species is threatened in all parts of its range and its numbers have become significantly reduced in the last few decades. For anthropology, on the other hand, local accounts of Komodo dragons – including the initial reports by Flores Islanders of 'land crocodiles' or lizards of 'unusual size' living exclusively on dry land (Ouwens 1912: 1) and inhabiting the Labuanbajo area and Komodo Island – have another, equally important significance. For they provide a clear demonstration of how non-western representations of a creature that may initially sound improbable to western ears can in fact be zoologically accurate and thoroughly grounded in empirical reality.

Wild cats

According to one possible interpretation, a putative wild cat on Flores – that is, a truly wild, not a feral, cat – may represent another instance of a known species showing up in a hitherto unsuspected place. Since wildmen, whatever their status in the past, are unlikely to be an empirical species of non-sapiens hominins that survives to the present day, and since Komodo dragons have, for some time, been well-known from Komodo Island and western Flores, for ethnozoology the identity of the wild cat, named separately from other cats as *ngo ngoe*, is in some ways more intriguing. My information on this cat derives from a variety of local accounts. These include interviews with eighteen men who, on at least one occasion, had seen specimens of *ngo ngoe*, mostly animals killed by hunters or caught in snares.

Besides *ngo ngoe*, Nage distinguish two other categories of local cats: *meo bo'a* (or the synonymous *meo ola*) and *meo witu*. *Meo bo'a* ('village cats') are domestic cats (*Felis catus*). *Meo witu* ('forest cats' or 'wild cats') are widely recognised among Nage as feral instances of *F. catus* and moreover as inter-breeding freely with domestic cats, which they generally resemble. At the same time, *meo witu* ('wild cat') is regularly used in a way that subsumes *ngo ngoe* as a sub-class (see Figure 2.1). Whether domestic or feral, *F. catus*, introduced to eastern Indonesia by Europeans at some time during the last four centuries and to some islands rather more recently, is, according to international zoologists (e.g. Monk et al. 1997) the only felid found on Flores. Existence of a third kind, the aforementioned *ngo ngoe*, would therefore contradict this view.

Nage describe *ngo ngoe* as currently rare, as extinct in some locations, and as probably headed for complete extinction. It is consistent with this, that of fourteen sighting reports in which dates were mentioned, five related to the 1960s or earlier and another five to the 1970s; hence just four sightings took

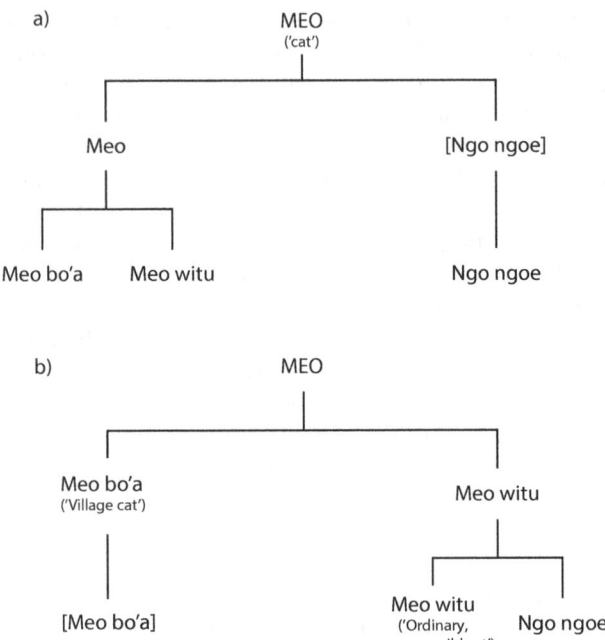

Figure 2.1 Two possible schemes of taxonomic relations among Nage felid categories

place after 1980, the most recent being in 2006 and 2009. Also, of eighteen informants who had seen *ngo ngoe,* mostly killed specimens, all but two had observed the animal only on a single occasion. A large majority of Nage in their forties or younger have never seen one, dead or alive. And, I should add, neither have I.

Nage regard *ngo ngoe* as a type of cat (*meo*), and some suggest they are simply large specimens of feral cats, or more specifically, large male specimens. Yet according to many people, including most eye-witnesses, *ngo ngoe* denotes a noticeably larger animal distinct from feral cats, which includes both males and females, and which is probably unable to mate with other felines. One eye-witness account concerned a specimen, killed during a hunt with dogs, which, although characteristically large, was female.[1] On the other hand, all other sighting reports concerned male animals, and Nage generally agree that the majority of *ngo ngoe* encountered within living memory have been males.

Nage who do not regard *ngo ngoe* as a kind distinct from other local felids, as well as those who do, describe the creatures as much bigger than feral or domestic cats. Most often they describe them as growing as large as a dog, or even a large dog. Of fourteen interviewees who mentioned size, seven compared the size to that of a young or year-old dog or a bitch, six to that of a 'large' or male dog, while one simply said a dog. Discussing a specimen they had observed some years previously, two men estimated the animal's length, from head to tail, as one metre and its height as half a metre; these dimensions one man then compared to a low table, which turned out to match his estimate almost exactly. Other men mentioned length estimates of about a metre, while yet another compared the size of an animal he had seen to that of his hunting bitch, a dog with a head and body length of 68cm and a shoulder height of 44cm. I was not able to measure the tail in this case, but by all accounts *ngo ngoe* tails are longer and straighter than those of other cats. Two men described specimens with tails as long as a man's hand and forearm, but in any case the tails are longer than those of domestic cats, which on Flores often have short and bent tails. Sometimes, Nage crop the tails of domestic cats, especially ones inclined to wander far from home which they infer were sired by feral toms. Several informants further described *ngo ngoe* tails as straighter than even a feral cat's, or as like a dog's. The legs are usually characterised as intermediate in size between that of a dog and of a domestic cat. One man described a live specimen he had encountered 'striding like a tiger' (Indonesian/ Malay *harimau*; referring apparently to tigers he had seen on television), by contrast to other cats whose strides are smaller and slower. Information on the head and muzzle of *ngo ngoe* was rather more consistent. According to a general representation, the shape and size of the body is like a dog but the head is that of a cat. Two men, however, described the muzzle as longer than in other cats, and hence more like a dog's, while another spoke of a relatively small muzzle in relation to a broad face and jaw. Several men also mentioned whiskers much longer than those of other cats. Information on the ears was similarly various: some thought *ngo ngoe* ears were proportionally larger or longer than those of other cats, but others did not.

Whereas domestic cat and feral cat pelage is recognised to vary, Nage usually describe the pelage of *ngo ngoe* as uniform, and typically as 'striped'. This they regularly compare to the fur of a tiger. A less often mentioned comparison is the striped hide of an immature wild pig. At the same time, two killed specimens of *ngo ngoe* were described as all 'black' (*mite*, which however can further denote dark shades of other colours). More generally as well, it is recognised that black exemplars can occur, although according to one informant these frequently show white or light-coloured spots on the tail. *Ngo ngoe* fur is characterised as coarser than the fur of domestic and feral cats, but smoother than a dog's. Referring to a particular specimen, one man described thick hair around the jaw.

Uttering a deep sound imitated as *'ngo ngoe'*, the cat's vocalisations, after which the animal may be onomatopoeically named, are also louder and deeper than other felids, and more resonant. These are most often heard at night, when the creature is not seen. According to one account, when treed by hunters (usually in daylight), *ngo ngoe* remain silent; in one opinion they will utter no sound even when subsequently brought down and savaged by hunting dogs.[2]

Ngo ngoe are stronger and more aggressive than feral cats and for this reason, and because of their greater ability to move rapidly in trees, they are far more difficult to hunt. Nage say that it requires at least four or five hunting dogs to kill one. According to one report, a lone hunter without dogs could find the animal standing its ground (unlike other wild cats, which would immediately flee), and he would then be in some danger. In the same vein, Nage claim that feral kittens (*ana meo witu*) can be raised and tamed but immature *ngo ngoe* cannot. Owing to their greater size and strength, *ngo ngoe* are further described as having a more varied diet than feral cats, being able to hunt larger animals such as immature deer, wild piglets, and Giant rats (*Papagomys armandvillei*) in addition to the smaller rodents and birds that are the mainstay of feral cats. This, however, appears to be mainly an inference, as I never found anyone who had actually observed *ngo ngoe* hunting the larger prey.

Owing to the creatures' greater size and aggressiveness, most Nage I questioned doubted whether *ngo ngoe* could mate with domestic cats; in this connection some also mentioned how male *ngo ngoe* – apparently despite their boldness in other settings – are wary of entering settlements where female domestic cats are found. At the same time, it was thought more likely that *ngo ngoe* can mate with other, smaller wild (feral) cats, although it seems that no one had actually observed such mating. Here it should be mentioned that Nage views on the possibility of successful mating do not invariably coincide with an assessment of two nominally distinct categories as belonging to different 'kinds' (a notion nowadays usually expressed with the Indonesian term *jenis*).[3] Thus, two experienced hunters, both of whom offered detailed accounts of *ngo ngoe*, claimed the creatures were a very distinct kind of wild cat that was 'original' (Indonesian *asli*) and could not possibly descend from feral felines, yet they did not rule out that they might be able to mate with feral cats. According to the one report I have on litter size, *ngo ngoe* produce smaller litters than other cats, comprising just two or three kittens. By contrast, both domestic and feral cats (*meo bo'a* and *meo witu*) have larger litters, as many as six or seven according to a usual estimate. International zoologists have specified litter size in domestic cats as between one and ten, with a mean between four and five (Deag et al. 2000: 25).[4]

Evidence for a Flores wild cat

In what follows, I consider how far Nage descriptions of *ngo ngoe* point to the presence on Flores of a truly wild felid that international zoologists have yet to document for the island. Afterwards, I discuss how far the same descriptions are consistent with the possibility that what Nage label as *ngo ngoe* are nothing more than very large specimens of feral cats – an interpretation which, as noted, is advanced by some (albeit probably a minority of) Nage themselves. Before discussing the two alternatives, however, attention should be given to another possibility, namely that *ngo ngoe* denotes a category that is wholly or largely imaginary.

Having listened to Nage talking about *ngo ngoe* during multiple visits over a period of thirty years, I can state categorically that nothing they say suggests they are anything other than empirical creatures. It is conceivable that they are partly imaginary, which is to say that their representation has its basis in ordinary felids but has been exaggerated and elaborated in a non-empirical direction (as, for example, in their claimed resemblance to 'tigers'). But what could motivate Nage to imagine, and moreover to name, a distinct kind of wild felid is quite unclear. In this connection, it is important that people describe *ngo ngoe* in a thoroughly prosaic or naturalistic way. They do not attribute to them any supernatural or otherwise implausible qualities. It is also worth noting that Nage regularly articulate a distinction of natural and supernatural beings, identifying the latter in their own language as entities that exemplify or manifest *nitu*, a general term for 'spirit, spiritual being'. Somewhat ironically, the mystery cats contrast in this respect with feral cats (*meo witu*) insofar as the latter are the explicit object of taboos on killing and consumption bound up with an association of *meo witu* with earth spirits (*nitu*; Forth 2007). In this context, *meo witu* could be understood as implicitly including *ngo ngoe*; but the point is that neither in this nor any other connection is *ngo ngoe* expressly named as a taboo category. In this respect, *ngo ngoe* are comparable to the similarly rare and disappearing Komodo dragons in north coastal Flores, since the giant reptiles also are scarcely subject to any fantastic representation – by contrast to all smaller kinds of lizards named and recognised by Florenese, which are indeed accorded manifestly non-empirical attributes (Forth 2010).

Considered as a hypothetical cultural construct, the image labelled *ngo ngoe* does not fulfil any obvious social or ideological function. Anyone trying to make this argument, therefore, must answer the question: if Nage have created a fictitious creature, why have they done so? Conceivably, a clue could be found in local comparisons of *ngo ngoe* to tigers. At present, tigers (*Panthera tigris*) still occur on Java and Sumatra and, until recently, on the island of Bali, which, like the two larger islands, falls west of Wallace's line and thus within in the Oriental biogeographic realm.[5] As Flores lies east of the line, and in the Australasian biogeographic realm, one should certainly not expect to find real tigers on the island. Nevertheless, it might just be argued that the category *ngo ngoe* has been culturally constructed to provide Florenese people with a local equivalent of western Indonesian tigers – perhaps as a way of representing their land as not so different from Java, or as a means of promoting a more diffuse equivalence with the Javanese. However, not only does this interpretation assume a conscious and special association of tigers with western Indonesians (and especially the numerically and politically dominant Javanese); it also requires that Nage knew of tigers prior to the development of the named category *ngo ngoe* and, moreover, that the socio-political contrasts between western and eastern Indonesians, hypothetically motivating a Florenese claim to local tigers, were equally recognised at the time. In fact, it is more likely that the tiger comparison is a twentieth-century addition to an established representation, or at least one subsequent to the category's emergence. Finally, it should be emphasised that Nage do not say *ngo ngoe* are tigers, or even that they are as big as tigers, but mostly that their markings resemble those of a tiger.

While one can be sure there are no tigers on Flores, there is, however, a wild felid that occurs on one eastern Indonesian island. This is the leopard cat (*Prionailurus bengalensis*). According to current evidence, eastern Indonesian leopard cats are found only on Lombok Island, immediately to the east of Bali, from where they may in fact have been introduced by humans (Monk et al. 1997). Yet if leopard cats were brought to Lombok, one could easily imagine their being later introduced to Sumbawa, the next large island further east, and from Sumbawa to the next, which is Flores.

One problem with the leopard cat hypothesis is the size of the *ngo ngoe*. Published sources describe leopard cats as not significantly larger than a domestic cat. What is more, Indonesian sub-species of *Prionailurus bengalensis*, a species with 'the broadest geographic distribution of all the small Asian cats' (Sunquist and Sunquist 2002: 226), are among the smallest. Descriptions of *ngo ngoe* as attaining the size of a dog, therefore, appear not to match leopard cat statistics. Yet it is important to note that Florenese dogs are themselves small, rarely growing larger than a medium-sized terrier or beagle and with a head to tail length usually no more than a metre – thus not larger than some estimates of *ngo ngoe*. In addition, published sources specifying *P. bengalensis* as the size of a domestic cat evidently refer to larger western house-cats, with an average head to tail length of 0.7 metres or more. Florenese village cats, by contrast, are typically far smaller than western cats, and like cats elsewhere in Indonesia and Malaysia, their tails are often 'short and bent' (Payne and Francis 2005: 290).

In addition to absolute size, body proportions also need to be considered. Recalling the long tail attributed to the *ngo ngoe*, it is noteworthy that leopard cat tails are about 40 to 50 per cent as long as the head and body. By comparison with the domestic cat, the head of the leopard cat is relatively small while the ears are 'moderately long' and 'rounded'. At the same time, leopard cats have proportionally longer legs than *Felis catus*, another feature consistent with a characterisation of the species as looking like a 'diminutive, more slender version of its namesake' (Sunquist and Sunquist 2002: 226) – which is to say, a leopard. As noted, Nage also speak of *ngo ngoe* legs as longer than a domestic cat's and, while others contradicted this, some even described the ears as relatively larger or longer. However that may be, relatively smaller heads combined with longer legs and larger ears could also contribute to an overall impression of a larger size, and one more comparable to canine dimensions than to Florenese specimens of *Felis catus*. Expressed another way, if *ngo ngoe* were as large as larger European specimens of *Felis catus*, and therefore of the size of a leopard cat (*Prionailurus bengalensis*), they could indeed appear considerably bigger than Florenese domestic cats, and even as large as smaller Florenese dogs.

Another factor seemingly counting against a hypothetical identification of *ngo ngoe* as leopard cats is the local comparison of *ngo ngoe* pelage with that of a tiger. As the English name suggests, leopard cats are more spotted than striped; also, rather than spots, darker markings may form rosettes (as, for example, in a jaguar's coat). In 2010, I showed electronic photographs of leopard cats to several Nage men of middle age or older who had closely observed at least one specimen of *ngo ngoe*. The exercise was not particularly productive (largely I think because some respondents had difficulty making out the images). However, of six who provided definite opinions, two judged certain pictures not to correspond to *ngo ngoe* pelage, because the leopard cat markings (rows of spots or striping) appeared to run from head to tail – as indeed they generally do in *Prionailurus bengalensis* – rather than from back to belly, a pattern they described as characterising both the large Flores wild cats and tigers. Nevertheless, the other four informants appeared not to recognise this difference; thus it would seem that many Nage simultaneously regard *ngo ngoe* as resembling both leopard cats and tigers.

Several considerations might explain this. First, Nage villagers know tigers only from media representations and, from insufficient familiarity, including lack of knowledge of the size of *Panthera tigris*, they may therefore be inclined to overgeneralise the tiger's similarity to better-known local felids. Second, leopard cat pelage can vary considerably even within sub-species; noteworthy here is Sterndale's (1884) claim that 'it is useless to lay down ... a very accurate description of the markings of this cat' (cited in Sunquist and Sunquist 2002: 226). Third, and relatedly, Indonesian leopard cat pelage, while otherwise spotted, does in fact show some striping. As Sunquist and Sunquist note (2002: 226), body markings of *Prionailurus bengalensis* sometimes 'coalesce to form lines', and the head is marked by two prominent dark stripes. Finally, the term Nage apply to striped pelage, *déto*, does not refer exclusively to stripes, or even to lines of spots. For it can denote other forms of distinct and regular markings, typically darker ones on a lighter background, as in both leopard cats and tigers, and also the brindled dogs common among Nage. Nage accordingly specify *ngo ngoe* pelage, which they regularly evaluate as 'striking' or 'handsome', mostly as *déto kune* or *déto to*, meaning dark 'stripes' on a 'yellowish' (*kune*) or 'reddish' (*to*) field. This characterisation would equally apply to the coat of a leopard cat, consisting of a yellowish or light copper-coloured ground marked with strongly contrasting dark stripes or spots.

Another feature possibly distinguishing *ngo ngoe* from *Prionailurus bengalensis* are the light or white under parts of the leopard cat. Responding to the aforementioned photographs, however, only one informant clearly specified this as a feature lacking in *ngo ngoe* he had seen. Lighter under parts, it is worth noting, have also been attributed to the 'large-eared cat' of Timor, another mystery felid I describe below. And they are also a feature of tigers.

Although recorded only once, another attribute of *ngo ngoe* consistent with *Prionailurus bengalensis* is litter size: leopard cats typically produce just two or three kittens (Sunquist and Sunquist 2002: 229), precisely the figure recorded for the Florenese wild cats. Other resemblances include the general representation of *ngo ngoe* as rarer and more wary of humans than feral cats (*meo witu*) and their greater association with wooded areas far from human habitations. Their reputed practice of occasionally entering villages and agricultural settlements to steal fowls is equally consistent with the leopard cat, which in Thailand is well known 'for attacking and killing domestic poultry' (Sunquist and Sunquist 2002: 228, citing Lekagul and McNeely 1977). As this should suggest, in the southern part of their range, which would include Indonesia, leopard cats are able to live near settlements, including villages, plantations, and areas of shifting cultivation. However, the cats are 'always associated with some type of forest cover', and even when they occur close to habitations, they are 'not often seen' owing to 'a liking for dense cover' (Sunquist and Sunquist 2002: 227).

Leopard cats are diurnal as well as nocturnal. According to local reports, *ngo ngoe* hunted by Nage have been chased and killed mostly in daylight, while several despatched at night were caught in snares set for other animals. Further suggestive of *Prionailurus bengalensis* are the Nage depiction of *ngo ngoe* as more aggressive than other cats, and their concomitant claim regarding the impossibility of taming immature animals. As the Sunquists note, of all Asian wild cats, leopard cats are reputedly 'the most difficult to tame ... and even month-old kittens may become unhandleable and intolerant of one another by one year of age' (2002: 229). In Sumatra, and especially in the province of Bengkulu, young leopard cats are captured and sold as pets (Sunquist and Sunquist 2002: 229, citing Santiapillai and Supraham 1985); but this presumably refers mostly to kittens acquired at a younger age.

The ability of *ngo ngoe* to move rapidly in trees – in which respect one eyewitness described a specimen as jumping between trees and branches faster than a long-tailed macaque (*Macaca fascicularis*) – similarly accords with *Prionailurus bengalensis*, a species characterised as 'agile climbers' and as 'quite arboreal in their habits' (Sunquist and Sunquist 2002: 227). Finally, it should be mentioned that leopard cats have been known to interbreed with domestic cats and to produce fertile, viable offspring (Sunquist and Sunquist 2002: 229). Although this contradicts the Nage opinion that *ngo ngoe* are too large and fierce to mate with other cats, especially village cats, it nevertheless suggests a sympatry with *Felis catus* which could illuminate divergent local opinions regarding the relation of *ngo ngoe* to the category *meo witu* (referring partly or entirely to feral cats).

Ngo ngoe as a large feral cat

I am not the first writer to consider the possibility of leopard cats on Flores. The missionary zoologist, lexicographer and Flores polymath, Jilis Verheijen, evidently entertained the same hypothesis. In his comprehensive dictionary of Manggarai, the language of western Flores, Verheijen provisionally identified *Prionailurus bengalensis* – or *Felis bengalensis* as the species was designated at the time – as the referent of a category of wild felid known to the Manggarai people as *énggo* (Verheijen 1967). To test this interpretation, Verheijen sent the carcass of an animal locally identified as an *énggo*, including the skull, to the Dutch Museum of Natural History in Leiden, where it was inspected by the mammalogist A. M. Husson.[6] According to Husson, who responded to Verheijen's query in a letter dated 8 November 1967, the cat's zygomatic arches differed somewhat from those of a specimen of a Dutch feral house-cat in the museum's possession. On the other hand, Husson noted how the zygomatic arch can be strongly influenced by diet, and furthermore that the length of the Flores cat's carnassial teeth indicated *Felis catus*. He therefore concluded that the creature was nothing other than a feral house-cat (Husson 1967).[7]

Quite apart from the lack of a confirmed specimen, other factors suggest that Nage *ngo ngoe* – apparently like Manggarai *énggo* – may refer to no more than especially large feral cats, as some Nage have themselves concluded. These may be feral cats displaying a striped (agouti) or tabby pelage, whose size has either been exaggerated or which have indeed grown unusually large.[8] Local suggestions that *ngo ngoe* denotes exclusively male cats apparently reflects the preponderance of males among specimens killed or observed and identified by this name. It is significant, then, that males of *Felis catus* can grow significantly larger than females, as seems especially to be the case among feral populations, where larger male size can be selected for in mating (Liberg et al. 2000: 137–138). Size is also a factor affecting hunting ability. As cats are typically unable to kill prey larger than themselves (Turner and Bateson 2000: 158), the larger the cat the larger will be its potential prey. By the same token, the availability of larger prey would select for larger cats, particularly in feral males whose ranges are typically larger than those of females. For Flores Island, there is no information on demographic or other factors that might favour large male size in gaining mates nor (other than informants' statements) is there any information on feral cat diets. However, it may be relevant that, in Australia, where regional habitats can quite closely resemble ones in eastern Indonesia, feral cats are known to reach a size considerably greater than domestic house-cats. One black-haired specimen, killed in June 2005 by a hunter in the Gippsland region of Victoria, had reputedly attained a head-to-tail length of over 1.7m.[9]

As a possible tendency especially of male feral cats, large size bears on the identity of *ngo ngoe* in another way. As noted, Nage village cats are small, and noticeably smaller than European specimens of *Felis catus* (which appear to provide the measure of species' size in the literature). However, it should further be considered whether the difference in size between these cats and *ngo ngoe* might not reflect the circumstance that cats seen inside Nage villages and houses are more often females, and that these mate with less often seen toms that spend most of their time near village peripheries or in forest. Adult male cats among both domestic and feral populations are considerably larger and can grow to one and a half times heavier than adult females (Fitzgerald and Turner 2000: 162). Also pertinent are differences in range behaviour in male and female cats. As noted, breeding males have much larger ranges than females, even three to four times larger, and these typically overlap with two or more female ranges. In addition, ranges of feral males can encompass the ranges of domestic females (Sunquist and Sunquist 2002: 106–107; Liberg et al. 2000: 126). It is important to recognise that domesticity or ferality in *Felis catus* is always a matter of degree, the two terms defining a continuum (Liberg et al. 2000: 120–121). In the Nage classification, therefore, *meo bo'a* ('village cat') may best be understood as a reference to cats attached to particular households but which also depend partly on hunting and scavenging – as can far better fed pet cats in the west. As suggested, these may be mostly females. *Meo witu* ('wild' or feral cats), on the other hand, probably refers largely to cats whose males mate with 'domestic' females, live mainly by hunting but also steal poultry and scavenge around village peripheries, and possibly have only relatively recently become independent of human settlements, or 'become feral'. Although a category itself classifiable as a type of 'wild cat' (*meo witu), ngo ngoe* would then denote populations that have long been feral and which mostly inhabit forests relatively distant from human habitations. This agrees with informant descriptions of *ngo ngoe* as living separately from other 'wild cats' (which are said to be afraid of them) in rock crevices and caves in more remote and densely wooded areas. Accordingly, while both *ngo ngoe* and smaller, feral cats are described as stealing fowls from settlements, *ngo ngoe* are known to do so far less often. In this view, *ngo ngoe* would also be reproductively self sustaining (or would be more so than other feral cats), and for reasons already noted would tend to a greater size than either 'domestic' toms or smaller 'wild' cats Nage classify as *meo witu*. As mentioned, Nage describe this last category, sometimes in contradistinction to *ngo ngoe*, as interbreeding with village cats, and specifically with village females.

While informant descriptions suggest the main difference between *ngo ngoe* and *meo witu*, or 'other wild cats', is size – the latter being mostly no larger than domestic cats – information on pelage differences is rather more variable. Ordinary wild cats (as they may be distinguished) are often described as possessing striped or spotted (*déto*) coats identical or similar to *ngo ngoe*. On the other hand, Nage say *meo witu* also occur in other colours, for example, all 'yellow', 'white', or 'black', or with mixed dark and light patches. Describing *meo witu* as generally the same size and colour as *meo bo'a* (village cats), one man even claimed that the only way to distinguish the former was by the contents of their stomachs; these, he noted, contain only wild foods, since only village cats are fed on rice and other cooked food. But according to a contrary view, wild cats with other than 'striped' pelage must be *meo coa* (*coa*, 'to flee, escape'), that is, village cats that have only recently left human settlements and have yet to become truly feral. Hence in this assessment, which appears to be empirically well founded, cats lose the various colours and markings characteristic of domestic cats, and revert to the basically tabby pelage of the ancestors of all domestic felines, the longer they have been 'wild'.

Reports of feral cats in other parts of the world indicate a preponderance of particular pelages, including tabby, agouti, orange, and tortoise-shell. In some rural areas, orange-coloured toms tend to be more aggressive – presumably as an effect of pleiotropy (the production by a gene of two or more unrelated effects in the phenotype) – and therefore better able to mate, so that this colour becomes more frequent (Mendl and Harcourt 2000: 52–54; see also Jones 1977). Since large male size is similarly advantageous in mating, one might further expect this colour to coincide with a larger body. But it is not clear that this colour-linked advantage applies everywhere, or only, to male cats with the orange-allele. On one subarctic island, a preponderant darker phenotype has been shown to reflect, in part and possibly as an effect of pleiotropy, better adaptation to a cold, wet, harsh environment (Van Aarde and Skinner 1981). Another study reports 'significant differences in body weights of cats between some localities and for one coat colour allele' (Jones and Horton 1984). Aggressiveness aside, a striped (agouti) or tabby pelage could conceivably confer a selective advantage as camouflage, particularly where wild cats are hunted by humans, as indeed they are on Flores.[10] Alternatively, the pelage could be linked pleiotropically with some other trait selecting for large size among Florenese feral cats.[11]

It may well be, therefore, that a preponderance of striped or tabby pelage in very big feral cats inhabiting areas at some distance from settlements has led to the impression of a distinct kind of large 'striped' cat thus comparable to a 'tiger'. Expressed another way, the preponderance may have given rise to a stereotype. To this may be added that, while tigers are nowadays known mostly from modern media, some image of tigers, conveyed either by pictures or word of mouth and relating to a large cat or cat-like animal existing on some other Indonesian islands, may very well have been available in eastern Indonesia for some time, and perhaps since the nineteenth century. This too could have influenced Nage cat classification, though as noted above, not to the extent of independently generating the named category *ngo ngoe*. An early familiarity with tigers may partly explain a curiosity of the Holle lists, word registers from numerous Indonesian languages based on a list of words in Dutch and Malay compiled by K. F. Holle (1829–1896), a member of a plantation family based in West Java (see Stokhof 1980–87). Among Holle's source terms is 'tiger' (Dutch *tijger*, Malay *harimau*), and terms supposedly corresponding to 'tiger' accordingly appear in some of the earliest word lists for Savunese and Rotinese, languages spoken to the south and southeast of Flores.[12] A word for 'tiger' (Malay *harimau*) also occurs in Verheijen's dictionaries of Manggarai (western Flores; 1967; 1970), here given as *matjang* (current transcription *macang*), an obvious borrowing from Javanese *macan* ('tiger').

Nage cat classification, and particularly the identity of *ngo ngoe*, is further illuminated by other comparative evidence. Mammal nomenclatures from the islands of Roti and Sulawesi suggest a tripartite classification of felids identical to that found among the Nage. Both Roti and Sulawesi, it should be noted, are, according to the current zoological consensus, islands which, like Flores, are inhabited by just one kind of cat, *Felis catus*. Besides *meò*, the general term for 'cat', Rotinese animal names thus include: *meò aek*, 'domestic or house cat' (*aek*, 'native, own'), *meò fuik*, 'wild cat', and *meò-asu*, 'cat as large as a dog, i.e. tiger' (Jonker 1908). *Meò-asu* also appears in the aforementioned Holle lists as the Rotinese word for 'tiger'. Since *asu* is a Rotinese term for 'dog', and appears with the same meaning in many other eastern Indonesian languages, Jonker's translation of *meò-asu* is curious, for not only are there no tigers anywhere near Roti but tigers are of course very much larger than dogs. On the other hand, the gloss 'cat as large as a dog', as well as some association with tigers, immediately – and quite remarkably – recalls Nage descriptions of *ngo ngoe*, and therefore suggests that, whatever the referent of the Nage term might be, Rotinese *meò-asu* quite probably refers to the same thing.[13]

Bolaang-Mongondow, a language of northern Sulawesi, similarly comprises three terms for felids: *pinggo'*, 'cat' (presumably referring primarily to domestic cats); *pinggo' lolog*, 'wild or feral cat', and *linggau*, 'forest cat' (Dutch *boschkat*). The lexicographer in this case, Dunnebier (1951), further remarks that the *pinggo' lolog* resembles the 'usual house cat' but differs from the *linggau*, which he describes as 'very shy'. From this characterisation, it is a reasonable inference that the latter animal is rarely encountered, by contrast to the *pinggo' loloq*, the feral cat whose resemblance to domestic cats recalls the resemblance Nage recognise between *meo* (or *meo bo'a*, 'village cats') and *meo witu*.

Wallace's tiger cat

Another kind of comparative evidence pertains not to folk nomenclature but to the development of western (and nowadays international) zoological knowledge of eastern Indonesia. In the original edition of *The Malay Archipelago* (1872 [1869]: 208), Alfred Wallace listed a very rare 'tiger cat', denominated as *Felis megalotis* (or 'large-eared cat'), as one of just seven land mammals to be found on Timor, the large island situated to the southeast of Flores and immediately east of the much smaller island of Roti. In addition, Wallace described the wild cat as peculiar to Timor and as existing only in the island's interior. Its 'nearest allies', he added, are to be found on the island of Java – in which connection he almost certainly alluded to the leopard cat (*Prionailurus bengalensis*). Interestingly, the name '*Felis megalotis*' is mentioned by Husson in his 1967 letter to Verheijen, in which he registered his opinion on the identity of the 'wild cat' specimen Verheijen sent him from Flores. Husson thus writes: 'I believe it is generally accepted at present that what was at one time described as *Felis megalotis* are feral house-cats, which occur throughout Indonesia.' Wallace's information was most likely taken either from Andrew Murray (1866) or the *British Museum Catalogue of Mammalia and Birds of New Guinea* (1854), from where Murray himself derived information on Timorese fauna. (Murray, it may be noted, describes the feline as further occurring on Sulawesi – the island which, according to Dunnebier's lexicon of Bolaang-Mongondow, would also appear to contain a distinctly named wild cat.) But the ultimate source for *Felis megalotis* is evidently the Dutch naturalist Salomon Müller, whose detailed description of the animal (in

Schlegel and Müller 1839–44) was apparently based on a specimen he himself had observed.

Müller's description is worth reviewing in detail. He describes the cat as a 'new kind' of felid which 'we found only on Timor'. The cat had 'fairly large' and 'spoon-shaped' ears which were much longer (or 'higher standing') than those of another cat he describes, *'Felis minuta'* – and presumably, therefore, also longer than those of a domestic cat. The tail, too, was 'noticeably longer' and had a flattened appearance, owing to the hairs being longest at the two sides. As regards pelage, the ground colour was generally a faded yellow, although more reddish-yellow on the back, while the under parts were more Isabelline (greyish yellow). On the head, shoulders, back legs and tail were black rings, giving the appearance of a dark marbling, while the lower side of the back legs showed black traverse stripes, and the forelegs reddish or dark stripes. Dark striping also occurred on the face, under the eyes and below the ears. The claws were light yellow, and the iris an orange-yellow. The body was 484mm, and the tail 300mm, thus giving a total length of 784mm.

This last figure, it should be noted, falls towards the high end of the normal range for both *Felis catus* and *Prionailurus bengalensis*. Allowing for variation in leopard cat pelage (which can display more striping or rosettes and can show a ground colour more yellowish or reddish), Müller's 'large-eared cat' – or 'tiger cat' as Wallace chose to call it – thus sounds remarkably like *Prionailurus bengalensis*. Noteworthy in this respect is the fact that leopard cats too are distinguished by ears larger and wider than those of *Felis catus*. And it is similarly interesting that Müller compares the animal's size to that of *Felis minuta*, once a synonym for *Prionailurus bengalensis*, especially in reference to the sub-species of leopard cat found in the Philippines.

At present, the stuffed and mounted type specimen of *Felis megalotis* – and therefore evidently the animal described by Müller – is in the possession of the Dutch National Museum of Natural History in Leiden.The specimen is recorded as female and, although Müller's description does not mention this feature either, the animal appears to have rather long legs. This may be significant, as longer legs are a characteristic distinguishing the leopard cat, *Prionailurus bengalensis*, from *Felis catus*.[14] Nevertheless, in spite of this and other correspondences with the former species, as Husson stated in his letter to Verheijen, the verdict of modern zoologists is that *Felis megalotis* was nothing other than a feral cat (*Felis catus*), and the binomial itself has long fallen out of zoological usage. Wallace himself evidently reached the same conclusion, for by the time his great book was published in a revised edition, *Felis megalotis* had disappeared from his list of Timorese mammals, whose number was thereby reduced from seven to six (1962 [1890]: 160). The reason for the deletion is suggested in Wallace's preface to the tenth edition (1890), where he states that, since his book first appeared, 'several naturalists have visited the [Malay] Archipelago', and that, in order to indicate the 'results of their researches', he has added footnotes 'whenever my facts or conclusions have been modified by later discoveries'. Wallace also refers to 'a few verbal alterations' made 'to correct any small errors or obscurities'. The 'tiger cat' of Timor, however, is not specified in this connection, nor is it mentioned anywhere else in the revised text.

Conclusion

It is quite possible that *ngo ngoe* denotes nothing more than especially large and striped or tabby feral specimens of *Felis catus*; yet the ethnographic evidence still leaves an element of doubt. Even if it could be shown that all Florenese cats were *Felis catus*, the existence of *ngo ngoe* as a distinctly named category does not in any way discredit local observation and classification of felids. All evidence suggests that Nage cat classification is solidly grounded in experience and that Nage are quite accurately observing and describing real animals albeit, in one case, of a variety which, according to their own evaluation, is nowadays rarely encountered. If *ngo ngoe* are simply feral cats, then many Nage folk zoologists – but not all of them – differ from international zoologists in giving these large specimens a distinct name and thus classifying them separately from other feral cats (*meo witu*). Otherwise expressed, their nomenclature reveals what ethnobiologists have called 'overdifferentiation' (Berlin 1973: 268). Yet the taxonomic 'error' with which some Nage might thereby be charged is entirely comparable to the assignment, by western naturalists and international zoologists, of specimens or populations of single felid species to different species – in which respect it should be noted that wide variation, especially in size (length, height, and weight), is typical of the Felidae in general (Sunquist and Sunquist 2002: 6). As clear an example as any of such western or international zoological 'error' is of course the assignment of what were probably no more than Timorese feral cats to the now discredited taxon, *Felis megalotis*.

   If, on the other hand, *ngo ngoe* does denote something other than a feral cat – and it is the international zoologists who are wrong and Nage folk zoologists who are right – the most likely candidate would obviously be the leopard cat (*Prionailurus bengalensis*). Since in Southeast Asia this cat belongs primarily to the Oriental biogeographic realm, if leopard cats do occur on Flores, then they almost certainly descend from specimens introduced by humans, as do monkeys, civets, deer, and several other mammalian species. The evidence of Nage folk zoological discourse is obviously not sufficient to confirm the leopard cat's occurrence on Flores. However, Nage descriptions are compelling, and do indeed suggest, at least as a hypothetical possibility, the presence on the island of a felid species other than *Felis catus*. Of course, anything short of a specimen is unlikely to convince international zoologists; nor should it. Yet the absence of physical evidence for a true Florenese wild cat does not mean that there is none to be discovered. After all, in recent years several previously undiscovered species of mammals (and in at least one instance a new genus) have come to light in Southeast Asia, including a Viverrid and a large rodent.[15] Since leopard cats are well known from western Indonesia and other parts of Asia, in this respect the Komodo dragon may provide a more apt comparison. Over the last hundred years, in all parts of Flores, reports by local people have consistently preceded documentation of the giant lizards by international zoologists (Forth 2010). Thus, one may ask, if local people in central and eastern Flores can be right about Komodo dragons, why cannot they be right about leopard cats as well?

To this, one might reasonably respond: they could be right but they are not so necessarily. But if the matter thus remains open, then Florenese folk mammalogy should provide justification for further zoological investigation in the field. Too often reports by non-western observers of creatures unrecognised by international zoology have been dismissed as unfounded and attributed to beliefs in imaginary or supernatural beings. Yet, as I have explained elsewhere, the people of Flores (to speak only of them) recognise an ontological distinction of 'natural' and 'supernatural' beings (Forth 2012a). The *ngo ngoe* clearly falls into what, for Florenese, is the 'natural' category, just as do local representations of Komodo dragons. The fact that the hominoid figures I have labelled 'wildmen' do so as well (Forth 2012a) possibly complicates the issue, but it by no means nullifies the distinction.

Rather than a reference to a 'mythical' creature subserving some indeterminable symbolic or narrative function, the Nage category *ngo ngoe* is a product of a classification of animal kinds grounded in experience of empirical species. The referent is either large feral cats of a certain appearance or an unrecorded felid species; any other interpretation, and particularly one which construes the reference as a supernatural or imaginary being, not only goes against both ethnographic and zoological evidence but fails to observe the principle of Occam's razor. Whether they ultimately refer to hitherto undocumented species or to specimens of known species erroneously construed as a different kind, cryptid categories like *ngo ngoe* provide anthropologists with a special challenge. But they also provide a special opportunity, a privileged context in which to explore and test the empirical validity of folk taxonomies. And insofar as they do not prove to be invalid, they can be significant for international zoology as well.

Acknowledgements

A shorter version of this paper was presented at the 32nd annual meeting of the Society of Ethnobiology, held in April 2009 at Tulane University, New Orleans, USA. I am grateful to several people in attendance for their comments. Thanks are also due to Dr Marie-Antoinette Willemsen, J. A. J. Verheijen's biographer, who kindly provided me with a copy of Husson's 1967 letter to Verheijen, and to Dr W. van der Molen, the librarian at the Royal Institute (KITLV) in Leiden, who supplied photographs of relevant pages from Schlegel and Müller (1839–44), an old and valuable source now scarce and difficult to obtain. Finally, I am grateful to Dr David W. Macdonald, Professor of Wildlife Conservation at Oxford University, for suggestions concerning the possible identity of non-domestic Florenese cats (pers. comm., 12 July 2009). It should go without saying that responsibility for any errors of fact or interpretation is mine alone.

Notes

1 Nage has a special term, *pode* or *'o'a pode*, for 'large male monkey' (cf. *'o'a*, dialectal *yo'a, ro'a*, 'monkey'). This, however, invariably denotes a male animal of a certain age, and unlike *ngo ngoe* in relation to *meo*, is never construed as designating an animal of a kind separate from other *'o'a*. The Manggarai language of western Flores appears to have more special terms for large males of wild species.

2 According to the same source, feral cats (*meo witu*) remain silent when treed but vocalize when savaged, whereas animals that have only very recently become feral (*meo coa*) will miaow when treed.

3 In other words, Nage can say that two zoologically closely related animals constitute different 'kinds' (*jenis*) and yet consider them able to mate. At the same time, they can judge two categories as belonging, at some level, to a single 'kind' and yet express doubts about the possibility of their interbreeding. Bearing on the question of folk conceptions approximating scientific conceptions of 'species', this is a complex issue that cannot be explored here (but see Forth 2016). It should be noted, however, that international biology, also, recognises members of different species, even genera, as being able to interbreed while regarding species united at some level of taxonomy (for example, as members of a single genus or family) as being incapable of hybridisation. Indeed, hybridisation of domestic cats (*Felis catus*) and Leopard cats (*Prionailurus bengalensis*), a phenomenon which is itself relevant to the present topic, provides an especially appropriate example of the former possibility.

4 Another difference informants mentioned between *ngo ngoe* and (other) *meo witu* concerns their flesh, which, when available, is freely consumed. One man described *ngo ngoe* flesh as tastier than that of other, smaller 'wild cats', which can have a putrid smell. Another compared the former to lamb or mutton. At the same time, the flesh of all 'wild' cats is reckoned to be tastier than the flesh of domestic cats; the latter is less regularly consumed, although not always clearly for this reason.

5 On the basis of both scratch marks and reputed sightings, it has been argued that tigers and leopards may still survive on Bali (World Wildlife Fund 1979).

6 According to records kindly provided by the museum's staff, the specimen, a complete cat, was collected in October 1958, in the central Manggarai region of Rahong. With other items in Verheijen's collection it was donated to the Leiden museum in 1967 (Steven van der Mije, pers. comm., February 2011).

7 As to why Verheijen might have thought a felid other than *F. catus* occurred in western Flores, it is noteworthy that *énggo* is distinct from Manggarai names for domestic cats, several of which, like Nage *meo*, are palpably onomatopoeic. In different dialects, these include *aong, ngéong, méong*, and *ngaong* (Verheijen 1967). Also locally described as onomatopoeic, Nage *ngo ngoe*, would appear more likely related to these than to *énggo*, but how far this supports an interpretation of the Nage category as a form of *Felis catus* is moot. Whether Manggarai people distinguish *énggo* from feral specimens of *Felis catus* as well as from domestic cats – as Nage do with the three terms *ngo ngoe, meo witu*, and *meo bo'a* – is not clear from the lexical materials provided in Verheijen's dictionaries (1967, 1970). Dialectal variants of *énggo* include *nggaro* and *nggaru* (1970; 115, s.v. *kutjing, kutjing hutan*). The last two terms are comparable to Bimanese *nggalu*, 'civet' (presumably the Palm civet, *Paradoxorus hermaphroditus*), a Viverrid in several respects similar to a cat which occurs throughout the eastern islands, and to Komodo Island *ngao nggalu*, 'wild cat' (Dutch *wilde kat*; Verheijen 1982).

8 'Agouti', a form of striping produced when each hair has alternate dark and light bands, describes the pelage of *Felis silvestris lybica/ornate*, the African and Asian wild ancestor of the domestic cat.

9 The estimate is based on a photograph taken by the hunter, Kurt Engel, and on the length of the tail (the only part of the animal that was kept), which is variously

reported as measuring 60 or 65cm. The animal was identified as a specimen of *Felis catus* from genetic testing performed at Monash University. The animal's sex appears not to have been recorded. Source: Darren Nash, 'Australia's new feral mega-cats', 4 March 2007, http://scienceblogs.com/tetrapodzoology/2007/03/04/a ustralias-new-feral-mega-cats/. According to the *Guinness Book of Records*, the largest domestic cat on record was 1.219m long.

10 A study of thirty-one feral cats in Scotland found that 74 per cent had 'striped tabby pelages'. However, this may partly reflect interbreeding with the European wildcat, *Felis silvestris silvestris* (Daniels et al. 2001).

11 Apart from humans, the only other known central Florense animals capable of preying on wild cats are feral dogs. Since many Florense dogs display a brindled coat, it is just conceivable that reports of large striped cats the size of dogs sometimes reflect encounters with feral canines. However, since *ngo ngoe* are generally represented as a kind of cat, and are moreover described as moving rapidly in trees, this would rule out wild dogs as the sole or principal source of the representation.

12 A list of Savunese words collected at the Savunese colony of Melolo in eastern Sumba, for example, gives 'tiger' as *meo ruba*. The term is clearly cognate with – and possibly a borrowing from – Sumbanese *meo rumba* ('wild cat'; *rumba* is 'grass, undergrowth, bush, scrub'), which in that language apparently denotes a feral cat. At present, Sumbanese themselves sometimes gloss *meo rumba* as 'tiger'. They further apply the name to a sort of hairy hominoid deployed as a bogey to frighten disobedient children (Forth 2012a). In the same vein, Wijngaarden (1896), who gives Savunese *meo ruba* simply as 'a kind of animal', additionally lists the expression *ne ne ke meo ruba*, which he describes as 'a saying used to scare children'.

13 In June 2010, after writing the foregoing, I had the opportunity to pursue this matter briefly with Rotinese people I met in Kupang. A man and a woman I questioned agreed that *meò-asu* referred to tigers (*harimau*), but said that the term could also be understood as a composite expression referring to 'cats and dogs' together. The woman further claimed that such *harimau*, or *meò-asu*, were still to be found on Roti and probably on Timor Island as well. The man agreed. Unfortunately, I was unable to clarify how these *meò-asu* might be related to the category *meò fuik* ('wild cat').

14 I am grateful to Professor Colin Groves of the Australian National University for alerting me to the existence of this specimen.

15 The new rodent genus is represented by *Laonastes aenigmamus*, the Laotian rock rat. The type specimen was discovered on sale as a comestible in a market in Laos in 2005.

References

Berlin, B. 1973. Folk systematics in relation to biological classification and nomenclature. *Annual Review of Ecology and Systematics*, 4, 259–271.

Ciofi, C. and Gibson, R. 2006. *Research report: A survey on the distribution and status of the Komodo monitor Varanus komodoensis in eastern Flores (Lesser Sundas, Indonesia)*. London: Zoological Society of London.

Daniels, M.J., Beaumont, M.A., Johnson, P.J., Balharry, D., Macdonald, D.W. and Barrat, E. 2001. Ecology and genetics of wild-living cats in the north-east of Scotland and the implications for the conservation of the wildcat. *Journal of Applied Ecology*, 38(1), 146–161.

Deag, J.M., Manning, A. and Lawrence, C.E. 2000. Factors influencing the mother-kitten relationship. In *The domestic cat: The biology of its behaviour*, edited by D.C. Turner and P. Bateson. 2nd edn. Cambridge: Cambridge University Press, 23–45.

De Jong, J.K. 1937. Een en ander over Varanus komodoensis Ouwens. *Natuurkundig Tijdschrift voor Nederlandsch-Indië*, 97(8), 173–208.

Downing, R. 1984. The search for cougars in the eastern United States. *Cryptozoology*, 3, 31–49.

Dunnebier, W. 1951. *Bolaang Mongondowsch-Nederlandsch woordenboek met Nederlandsch-Bolaang Mongondowsch register*. 's-Gravenhage, the Netherlands: Martinus Nijhoff.

Fitzgerald, B.M. and Turner, D.C. 2000. Hunting behaviour of domestic cats and their impact on prey populations. In *The domestic cat: The biology of its behaviour*, edited by D.C. Turner and P. Bateson. 2nd edn. Cambridge: Cambridge University Press, 151–175.

Forth, G. 2004. *Nage birds: Classification and symbolism among an eastern Indonesian people*. London and New York: Routledge.

Forth, G. 2007. Can animals break taboos?: Applications of 'taboo' among the Nage of eastern Indonesia. *Oceania*, 77(2), 215–231.

Forth, G. 2010. Folk knowledge and distribution of the Komodo dragon (*Varanus komodoensis*) on Flores Island. *Journal of Ethnobiology*, 30(2), 293–311.

Forth, G. 2012a. *Images of the wildman in Southeast Asia: An anthropological perspective*. London and New York: Routledge.

Forth, G. 2012b. Are hairy hominoids worth looking for? Views from ethnobiology and palaeoanthropology. *Anthropology Today*, 28(2), 13–16.

Forth, G. 2016. *Why the porcupine is not a bird: Explorations in the folk zoology of an eastern Indonesian people*. Toronto: University of Toronto Press.

Hurn, S. 2009. Here be dragons? No, big cats!: Predator symbolism in rural West Wales. *Anthropology Today*, 25(1), 6–11.

Husson, A.M. 1967. Letter of 8 November to Father J.A.J. Verheijen SVD. (Unpublished).

Jones, E. 1977. Ecology of the feral cat on Macquarie Island. *Australian Wildlife Research*, 4, 249–262.

Jones, E. and Horton, B.J. 1984. Gene frequencies and body weights of feral cats, *Felis catus* (L.), from five Australian localities and from Macquarie Island. *Australian Journal of Zoology*, 32(2), 231–237.

Jonker, J.C.G. 1908. *Rottineesch-Hollandsch woordenboek*. Leiden: E.J. Brill.

Lekagul, B. and McNeely, J.A. 1977. *Mammals of Thailand*. Bangkok: Association for the Conservation of Wildlife.

Liberg, O., Sandell, M., Pontier, D. and Natoli, E. 2000. Density, spatial organisation and reproductive tactics in the domestic cat and other felids. In *The domestic cat: The biology of its behaviour*, edited by D.C. Turner and P. Bateson. 2nd edn. Cambridge: Cambridge University Press, 119–147.

Mendl, M. and Harcourt, R. 2000. Individuality in the domestic cat: origins, development and stability. In *The domestic cat: The biology of its behaviour*, edited by D.C. Turner and P. Bateson. 2nd edn. Cambridge: Cambridge University Press, 47–64.

Monk, K., De Fretes, Y. and Reksodiharjo-Lilley, G. 1997. *The ecology of Nusa Tenggara and Maluku*. The Ecology of Indonesia Series, vol. V. Hong Kong: Periplus Editions.

Ouwens, P.A. 1912. On the large Varanus species from the island of Komodo. *Bulletin du Jardin botanique de Buitenzorg*, 2(6), 1–3.

Payne, J. and Francis, C.M. 2005. *A field guide to the mammals of Borneo*. Kota Kinabalu, Malaysia: The Sabah Society.

Santiapilli, C. and Supraham, H. 1985. On the status of the leopard cat (*Felis bengalensis*) in Sumatra. *Tigerpaper*, 12, 8–13.

Schlegel, H. and Müller, S. 1839–44. *Verhandelingen over de natuurlijke geschiedenis der Nederlandsche Overzeesche Bezittingen, door de leden der Natuurkundige Commissie in Indië en andere schrijvers; uitgeven op last van den Koning door C.J. Temminck.* Vol. 2, Zoologie. Leiden, the Netherlands: S. en J. Luchtmans en C.C. van der Hoek.

Sterndale, R.A. 1884. *Natural history of the mammalia of India and Ceylon.* Calcutta: Thacker and Spink.

Stokhof, W.A.L. (ed.) 1980–87. *Holle lists: Vocabularies in languages of Indonesia.* 11 vols. Pacific Linguistics Series D. Canberra: Department of Linguistics, Research School of Pacific and Asian Studies, Australian National University.

Sunquist, M. and Sunquist, F. 2002. *Wild cats of the world.* Chicago and London: University of Chicago Press.

Turner, D.C. and Bateson, P. (eds) 2000. *The domestic cat: The biology of its behaviour.* 2nd edn. Cambridge: Cambridge University Press.

Van Aarde, R.J. and Skinner, J.D. 1981. The feral cat population at Marion Island: colonization, characteristics, and control. *Comité National Francais Recherches Antarctiques*, 51, 281–288.

Van Suchtelen, B.C.C.M.M. 1921. *Endeh (Flores).* Mededeelingen van het Bureau voor de Bestuurszaken der Buitengewesten, bewerkt door het Encylopaedisch Bureau, Aflevering 26. Weltevreden: Papyrus.

Verheijen, J.A.J. 1967. *Kamus Manggarai I: Manggarai-Indonesia.* 's-Gravenhage: Martinus Nijhoff.

Verheijen, J.A.J. 1970. *Kamus Manggarai II: Indonesia-Manggarai.* 's-Gravenhage: Martinus Nijhoff.

Verheijen, J.A.J. 1982. *Komodo: Het eiland, het volk en de taal.* Verhandelingen van het Koninklijk Instituut voor Taal-, Land-, en Volkenkunde 96. The Hague: Martinus Nijhoff.

Wallace, A.R. 1872 (1869). *The Malay archipelago: The land of the orang-utan and the bird of paradise: A narrative of travel, with studies of man and nature.* London: Macmillan.

Wallace, A.R. 1962 (1890). *The Malay archipelago: The land of the orang-utan and the bird of paradise: A narrative of travel, with studies of man and nature.* 10th edn. New York: Dover.

Wijngaarden, J.K. 1896. *Sawuneesche woordenlijst.* 's-Gravenhage: Martinus Nijhoff.

World Wildlife Fund. 1979. *Report on the presence of larger cats on Bali Island and a survey on the Rothschild's starling (Leucopsar rothschildi, Streseman 1912).* Bandung, Indonesia: World Wildlife Fund.

3 Cryptids and credulity

The Zanzibar leopard and other imaginary beings

Martin T. Walsh and Helle V. Goldman

The title of this book would justify the inclusion of Prince Hamlet, of the point, of the line, of the surface, of n-dimensional hyperplanes and hypervolumes, of all generic terms, and perhaps of each one of us and of the godhead. In brief, the sum of all things – the universe. We have limited ourselves, however, to what is immediately suggested by the words 'imaginary beings'; we have compiled a handbook of the strange creatures conceived through time and space by the human imagination.

(Jorge Luis Borges and Margarita Guerrero, *The Book of Imaginary Beings*, translated by Norman Thomas di Giovanni (1969: 12))

Introduction

Borges and Guerrero's playful collaboration, referred to above, takes the form of a parody of medieval bestiaries, and comprises a series of descriptions of 'imaginary beings', some of them imagined by the authors themselves and mixing real and sham erudition in Borges' characteristic style (di Giovanni 2003: 133). It was originally published in 1957, anticipating the appearance in English of Bernard Heuvelmans' *On the Track of Unknown Animals* (1958), the *locus classicus* of cryptozoology and model for numerous modern bestiaries.[1] Cryptozoology, literally 'the science of hidden animals', has been defined as 'the study of the evidence for animals that are undescribed by science' (Eberhart 2002: xlvii). 'Undescribed' is also unrecognised, and therein lies the contradiction that has led some critics to dismiss cryptozoology as a pseudoscience (for example Simpson 1984: 1), and that exposes its origin in the compilations of 'fantastic zoology' that are spoofed by Borges and his co-author(s) (Dendle 2006).

There is nothing intrinsically unscientific about searching for new (undescribed) and old (extinct) species, and one cryptozoologist has defined the discipline concisely as 'a targeted-search methodology for zoological discovery', noting that it is only one of a number of possible means to achieving this end (Arment 2004: 9). But all too often cryptozoologists' desire to find hidden species and identify the imaginary as real leads them to downplay the negative evidence that carries more weight with conventional zoologists and ethnozoologists (Simpson 1984: 12–14; Meurger 1988: 11–24). Although there is also a strong tradition of debunking fakes and false claims within cryptozoology, it has failed to establish itself as an academic discipline (Coleman 2002: xxxiii), while the professional association founded by Heuvelmans and colleagues – the International Society of Cryptozoology – has long since been defunct and its journal (*Cryptozoology*) extinct (Eberhart 2002: xxvii; see also Turner, this volume).

Our epigraph highlights a boundary problem that cryptozoologists have also struggled with: what kinds of phenomena or imaginary being fall within their remit? The subjects of cryptozoology are generally now referred to as cryptids, on one definition 'the alleged animals that cryptozoologists study' (Eberhart 2002: xxiii, also xlvii). Some restrict this to non-microscopic creatures they consider most likely to be discovered to be living species; others include historical and contemporary entities that are more obviously mythical (compare Greenwell 1985; Arment 2004: 11–12, 16–18). In his encyclopaedia Eberhart takes the broader view, and lists ten categories that most of the 'mystery animals' in his compilation fall into:

1. *Distribution anomalies* [...]
2. *Undescribed, unusual, or outsize variations of known species* [...]
3. *Survivals of recently extinct species* [...]
4. *Survivals of species known only from the fossil record into modern times* [...]
5. *Survivals of species known only from the fossil record into historical times* [...]
6. *Animals not known from the fossil record but related to known species* [...]
7. *Animals not known from the fossil record or bearing a clear relationship to known species* [...]
8. *Mythical animals with a zoological basis* [...]
9. *Seemingly paranormal or supernatural entities with some animal-like characteristics* [...]
10. *Known hoaxes or probable misidentifications* [...]

(2002: xxiii–xxiv)[2]

The subject of our chapter is the analysis of narratives and statements about an animal, the Zanzibar leopard (*Panthera pardus adersi*), which might be included in Eberhart's third category, given that it has been declared by some authorities to be extinct, though many Zanzibaris remain convinced of its continued existence. The classic example of a carnivore in this category is the Thylacine or Tasmanian tiger (*Thylacinus cynocephalus*), and in a later section we will explore some of the differences and significant parallels between the two cases. Unlike its marsupial analogue, the demise of the Zanzibar leopard has been so recent that zoologists cannot be sure that no individuals survive. In this respect its cryptozoological (and ontological) status remains uncertain.

However, as an anthropological analysis of indigenous and other narratives about the leopard shows, it can be described under more than one of Eberhart's headings: as the undescribed (and unusual) variation of a known species, as a mythical animal with a zoological basis, and perhaps even as a supernatural entity with animal-like characteristics. Whereas popular cryptozoology typically reduces the investigation of imaginary beings to a single dimension ('do they exist or not?'), we argue that only careful anthropological and ethnozoological research can unravel the complexity of cases like that of the Zanzibar leopard and other so-called cryptids. And, as this introduction has implied, crypto-zoology itself can also be analysed anthropologically, a subject we will return to briefly in our conclusion.

The Zanzibar leopard in scientific discourse

The Zanzibar leopard, *Panthera pardus adersi* (Pocock 1932), is an elusive subspecies of leopard endemic to Unguja, the principal island of the Zanzibar archipelago, off the coast of mainland Tanzania. This, at least, is how we understood its status in 1996, when we began our joint research on local knowledge about the leopard and the anthropogenic threats to its survival (Walsh 1996; Goldman and Walsh 1997).[3] As we shall see, this statement now has to be qualified in the light of global studies of leopard genetics and new information about the likely impacts of hunting and habitat destruction on the local leopard population.

Although a number of visitors to Zanzibar in the nineteenth century noted the presence of leopards on Unguja, they did not become subjects of scientific investigation until the second decade of the twentieth century.[4] The first properly curated specimen was collected by the Zanzibar Protectorate's Economic Biologist, William Mansfield-Aders, and sent to the British Museum in 1919. The mammal specialist at the museum, Reginald Pocock, noted that this and a second specimen were morphologically distinct from continental African leopards (*Panthera pardus pardus*), being significantly smaller in size and with a pelage of rosettes so densely packed together that they appeared to disintegrate into spots. In a subsequent major review of Africa's leopards he described the Zanzibar leopard as a separate subspecies and nominated the skull and skin collected by Aders as the holotype (Pocock 1932: 563).

Apart from collecting specimens, British officials and other foreign residents made little attempt to study the Zanzibar leopard during the lifetime of the British Protectorate. There are six known specimens in museums, all deriving from the period before the Second World War: three in the Natural History (formerly British) Museum in London, two in the Harvard Museum of Comparative Zoology in Cambridge, Massachusetts, and a mounted specimen in the Zanzibar (Peace Memorial) Museum (for details see Walsh and Gold-man 2008). But very little was written about the leopard either before or after Pocock's description, and there are only scattered references to its natural history in the literature of this time (for example Mansfield-Aders 1920: 329). The British authorities were rather more interested in the protection of the leopard for its presumed economic value in keeping bushpigs (*Potamochoerus larvatus*) and other agricultural 'vermin' under control. This position brought them into conflict with villagers who felt strongly that leopards themselves must be controlled because of their attacks on people (small children in particular) and livestock, widely believed to be the work of witches. Despite attempts by the British to restrict leopard-hunting and trapping, rural Zanzibaris continued to take the law into their own hands and deal with leopards (and witches) as they saw fit (Walsh and Goldman 2007: 1138–1148).

After the Zanzibar Revolution in 1964 local practice became national policy and statutory protection of the leopard was ignored. A nationwide campaign of witch-finding and leopard-killing ensued, with tacit support from the revolutionary government. When this petered out in the early 1970s, leopard-hunting continued as part of a government-subsidised campaign against animals considered to be vermin, and was still officially sanctioned when we began our joint research in 1996. Meanwhile, the Zanzibar leopard began to attract outside attention for the first time since the publication of Pocock's description in 1932. Although subsequent authors had followed Pocock (Allen 1939: 245; Moreau and Pakenham 1941: 119; Swynnerton and Hayman 1951: 335), Czech zoologist Luděk Dobroruka (1964) questioned his decision to name the Zanzibar leopard as a separate subspecies, and argued instead that it belonged to the eastern African form *Panthera pardus suahelica*. The wildlife artist and researcher Jonathan Kingdon, however, reiterated the distinctiveness of its pelage and reinstated it as an island endemic and subspecies in contrast to the African leopard, *P. p. pardus*, a category that subsumes the continental races (including *suahelica*) recognised by Pocock and later 'splitters' (1977: 351; 1989: 45). In his compilation of information on all of Zanzibar's mammals, R. H. W. Pakenham, a former British administrator who had helped send one of the Harvard specimens, agreed with Kingdon and explicitly rejected Dobroruka's position (1984: 48).

Little new information came from Unguja itself until the early 1980s, when Tanzanian research students were working on the island. Issai Swai provided an up-to-date description of the status and distribution of the Zanzibar leopard in his thesis on wildlife on the island (1983: 19–20, 48, 52–53). He also described what are still the only reported sightings of the leopard by researchers, one by himself at Mapopwe, another by colobus researcher Fatina Omari (Mturi) at Jozani (1983: 53).[5] Unfortunately no one followed up with further fieldwork on the leopard, and little more was written about it until the 1990s, when donor agencies and NGOs promoting a conservation agenda began working closely with the Government of Zanzibar. Tony Archer, a Kenyan wildlife consultant working in Zanzibar, played an important role in bringing the Zanzibar leopard to wider attention (Archer et al. 1991: 65; 1994: 2, 17). Archer encouraged and advised many of the researchers who followed him, ourselves included. In 1994–95 three students wrote papers about the Zanzibar leopard, two of them young Americans taking part in an international study programme (Marshall 1994; Selkow 1995), and one a Zanzibari attached to the College of African Wildlife Management at Mweka (Khamis 1995). Their studies provided useful new information based on interviews with rural Zanzibaris, and led directly to our own survey of practices and beliefs threatening the Zanzibar leopard (Walsh 1996; Goldman and Walsh 1997), which was commissioned by the CARE-funded Jozani–Chwaka Bay Conservation Project (JCBCP) and undertaken in July 1996.

One of the issues that we investigated was the likelihood of the Zanzibar leopard's survival. While a number of authors had already presumed it to be extinct (Smithers 1971: 3; Hes 1991: 165; Miththapala et al. 1996: 1126; Nowell and Jackson 1996: 27, fig. 6), we concluded that although it was critically endangered, a small population of leopards probably remained on Unguja. We argued that further research and a conservation programme were urgently required to stave off the imminent extinction of the Zanzibar leopard, and suggested a number of ways in which this might be approached (Goldman and Walsh 1997: iii–iv). Following our study, Goldman began to draft a leopard conservation programme for JCBCP, and South African wildlife researchers Chris and Tilde Stuart were recruited to determine the leopard's current status and distribution. However, they failed to find any firm evidence for the presence of leopards, and concluded that the Zanzibar leopard was either extinct or very nearly so (Anonymous 1997; Stuart and Stuart 1997a: 2–4; 1997b: 1; 1998: 35–37). Discouraged by this, JCBCP managers dropped the draft conservation programme and further research on the leopard was abandoned. Although the Zanzibar leopard continues to be reported as extinct (for example Uphyrkina et al. 2001: 2618; Kingdon 2004: 174; Finke 2010: 136), we have argued that the evidence for this is equivocal (Goldman and Walsh 2002). The Zanzibar government and Tanzanian researchers assume its continuing existence (for example Nahonyo et al. 2002: 48–49, 57, 84, 86–87; Ame 2003), and the WWF's *Strategic Framework for Conservation 2005–2025* for the Eastern Africa Coastal Forests Ecoregion lists knowledge of its population status as one of its indicators to be achieved by 2015 (Mugo 2006: 16).[6]

While there is uncertainty over the continued existence of the Zanzibar leopard, studies of leopard phylogenetics have cast renewed doubt on its classification as a separate subspecies. Whereas Pocock recognised twenty-seven trinomial subspecies of *P. pardus*, twelve of them African, genetic research has tentatively reduced the global total to a minimum of nine subspecies, including only one African subspecies, *P. p. pardus* (Miththapala et. al. 1996; Uphyrkina et al. 2001; O'Brien and Johnson 2005: 416–417). The Zanzibar leopard has not been included in these genetic analyses – indeed it has been presumed extinct by the researchers – and as a result is now generally excluded from subclassifications of the world's leopards. However, the confirmation of two insular subspecies – the Sri Lankan leopard (*P. p. kotiya*) and the Javan leopard (*P. p. melas*) – through genetic work suggests the possibility that the Zanzibar leopard is similarly distinct from its mainland cousins, and that this hypothesis should at least be tested.[7] We are participating in a project to do just this, and determine its phylogenetic position. Although a good case can be made for directing conservation towards separate populations regardless of their subspecific status (Jackson 2000), decisions about taxonomic status can influence the perceived significance of extermination. The Zanzibar leopard has undoubtedly suffered in this regard, just as proposals for its conservation have been undermined by premature pronouncements of its demise.

These debates about the status of the Zanzibar leopard remind us that science comprises disputed hypotheses and competing narratives, and that this is true whether one views research from a broadly positivist or, by contrast, a social-constructivist perspective. It should be clear from the preceding account and our own participation in research and publication about the Zanzibar leopard that we ourselves adopt a position of scientific realism, though we are also keenly aware of the role that cultural and subcultural constructs can play in research. This approach forms the basis of our discussion of local narratives about the Zanzibar leopard, the subject of the sections which follow.

Ethnotaxonomies of the Zanzibar leopard

One of the most common strategies of cryptozoological research is to speculate on and seek to verify the identity of animals recognised ethnozoologically but not (yet) matched with known species or subspecies (Arment 2004: 56–73; see also Forth, this volume). Many town dwellers are unaware of the past or present presence of leopards on Unguja, though they do know about leopards (Swahili: *chui*) more generally through film, the media, folk tales, the name of a local football team, and the sale of leopard-print cloths. Zanzibaris in the rural areas or with a rural background are more likely to have heard of the Zanzibar leopard, especially those living in the villages on the coral rag areas where leopards were most common in the twentieth century (see the distribution maps in Swai 1983: 48; Goldman and Walsh 2002). Hunters and others with a more intimate knowledge of the island's leopards describe different varieties, but disagree on the names and definitions of these. As we shall see, it is not at all clear what their individual and in some cases idiosyncratic taxonomies refer to, and we are left with the equivalent of a cryptozoological puzzle.

The terminology of leopard varieties is specialised knowledge that cannot be found in any of the published Swahili dictionaries, including lexical collections of the dialects of rural Unguja. Pakenham (1984: 69) recorded *chui mwanzi* as a possible local name for the leopard in Mangapwani, north of Zanzibar town. But the existence of different named varieties didn't become apparent until Marshall (1994: 9) reported that all of his interviewees believed that there were two different varieties of leopard, *kisutu* and *konge*. Selkow (1995: 13) added a third type called *mwanzi*, this being an abbreviated version of the binomial recorded by Pakenham. He also reported the use of the names *kichigi* and *uwanda* as alternatives for *kisutu* and *konge* respectively, and speculated that the *konge/uwanda* variety was a 'black panther' or melanistic form of the leopard. A similar three-term (*kisutu, konge, muanzi* [sic]) and three-variety classification was given by the Zanzibari forestry officer Khamis (1995: 7). Although Selkow recognised that there was no clear consensus on the leopard's names or morphological description, the simpler classification reported by Marshall has since been reproduced in tourist guides as a natural fact (for example Else and Tyrrell 2003: 57; McIntyre and McIntyre 2009: 54).

As it happens, the ethnotaxonomy of the Zanzibar leopard is even more complex than reported by Selkow. In July 1996 we elicited twenty-one terms for different varieties from twenty-nine interviewees, most of them hunters who claimed some familiarity with leopards.[8] A further four informants described or knew of varieties but could not name them, two denied the existence of different types, and three (including a school teacher in Nungwi) admitted ignorance. Some of the terms are variants of the same nominal root or abbreviations of binomials of the form [*chui* (leopard) + qualifier], in which only the qualifier is retained (for example *unyasi* < *chui unyasi*). Interpreting these as equivalents, this still leaves fifteen different core terms. As Table 3.1 shows, however, there is considerable variation in the knowledge and application of these terms. Most interviewees recognised only two or three leopard varieties, but no two informants agreed entirely on the naming and description of these. Subsequent field research (for example by Goldman in 2003) has confirmed this pattern without adding to the list of names we recorded in 1996.

The existence of such divergent taxonomies mirrors a general socio-linguistic pattern on Unguja of local variation in the terminologies for plants and animals, the names of which sometimes vary from one village to the next. It presumably also reflects the fact that knowledge about the Zanzibar leopard is restricted, and has become increasingly so as fewer people have had direct experience of it (Walsh 1997; 2000). Under these circumstances it is not surprising that individual classifications should have become muddled, borrowing names and perhaps also categories from one another, as well as importing them from other taxonomies (for example of bushpigs) and vocabularies (for example the adoption of euphemisms as names for particular varieties). Ignoring the actual terms used, however, it is clear that they represent similar and overlapping sets of discrimination based primarily on perceived differences in the size, build, coat pattern and colour of leopards. The next question to ask is whether these reflect real differences in the morphology of individual leopards.

Knowledge of leopards elsewhere would lead us to expect that the principal observable differences are between animals of different ages and sex. Adult male leopards are generally larger than their female counterparts. We are not aware of studies that have documented male–female differences in pelage, which can vary by age – in older animals it can become paler in colour (Sunquist and Sunquist 2002: 320) – as well as look different according to the condition an individual is in, the amount of dust on its coat, and the light in which it is viewed. It is also the case that no two leopards have exactly the same coat pattern, and that there can be considerable variation in a single population as well as the presence of aberrant individuals (Kingdon 1977: 350; Sunquist and Sunquist 2002: 319–320). Melanistic leopards are relatively rare in East Africa, and there is no evidence in Zanzibaris' own descriptions to suggest that these have ever been found on Unguja, contrary to Selkow's supposition (see above). Unfortunately the six Zanzibar leopard specimens in museums provide scant information with which

Table 3.1 Names for Zanzibar leopard varieties elicited in July 1996

Swahili name	Literal translation (if transparent)	Informant(s) (by residence)	Description given (salient features)
bete	'Dwarf (leopard)'	Chwaka	Short, robust
kibete	'Small dwarf (leopard)'	Zanzibar town	Small, a synonym for *kichigi*
bungala	'Bengali (-like leopard)' (also applied to large bushpigs and varieties of banana, rice, and sugar-cane)	Charawe	Large, long, with glossy coat
chui asili	'Original (typical) leopard'	Kizimkazi	Large, yellowish, fiercer than *chui uwanda*
chui uwanda	'Bushland leopard'	Kizimkazi	Small, rufous
		Uzi 1	Small and shorter than *kisutu*, rufous, with small red patches
futizi		Jambiani	More gracile than *kisutu*, but less so than *konge*
kariuki		Charawe	Small or medium-sized, with rufous patches
keke	'Disquiet, fear'; euphemism for leopards in general according to other informants in Kitogani and Makunduchi (1 and 2), and Chum (1994: 29)	Muungoni	Large, the largest of three (otherwise unnamed) varieties
kichigi	'Mannikin, small bird sp. (-like leopard)' (also applied to small bushpigs)	Makunduchi 1	Short, rufous (but no white)
		Makunduchi 2	Small, rufous
		Unguja Ukuu	Small, short, black and white (but no yellow), very fierce
		Zanzibar town	Small, a synonym for *kibete*

Table 3.1 (continued)

Swahili name	Literal translation (if transparent)	Informant(s) (by residence)	Description given (salient features)
koko	'Wild (leopard)'	Paje	Small, shorter than *mkonge*
konge	'Sisal leaf (-like leopard)'	Jambiani	Tall, slender
		Makunduchi 1	Long, red and black with a little white
		Makunduchi 2	Two kinds: (1) black with big spots; (2) white with small black spots
		Muungoni	Long, with spots which reduce the white background, and a big head
		Kitogani 2	Long and tall (hence the name), black and white
konga	Variant of *konge*	Muyuni	Large, with larger and more numerous spots than *unyasi*; the most common type
mkonge	'Sisal plant (-like leopard)' (also applied to large bushpigs)	Kitogani 1	Long, yellow, with some white
		Paje	Long
		Uzi 2	Larger than both *kisutu* and *unyasi*, coloured like sisal leaves
		Zanzibar town	Large
mwanzi	'Bamboo plant (-like leopard)'	Chwaka	Long, slender, yellow, with large black spots
		Ndudu	Long, coloured like *kisutu* cloth (pointing to a green and yellow example)
		Pete 1	With smaller marks than *kisutu*
		Zanzibar town	Richer yellow than *kichigi* and *mkonge*

Swahili name	Literal translation (if transparent)	Informant(s) (by residence)	Description given (salient features)
ngawa	'African civet (-like leopard)'	Ndudu	Patterned and coloured like the African civet (*Civettictis civetta*)
shambi-shambi	Cf. *shambi*, 'a gazelle' (Harries 1959: 68)	Pete 3	Similar to the African civet
shwambu	Euphemism for leopards in general according to other informants in Pete (1 and 2), Kitogani, and Jambiani	Pete 3	Very long, with very black and very white spots, and larger areas of white
sutu		Kitogani 2	= *kisutu*
kisutu	Diminutive of *sutu*; also the name of a kind of *kanga*, patterned cotton cloth, worn by women: the *kisutu* is distinguished by its wide border and is typically rufous, black and white in colour	Dimani	Same size as other kinds, but a different colour, with black and yellow spots
		Jambiani	Large, robust
		Jozani	Black, coloured like *ngawa*, the African civet
		Kitogani 2	Short, yellow and black, a better colour than other varieties and more marketable
		Muungoni	With more white than *konge*
		Pete 1	With black stripes like a zebra, large splotches or spots
		Pete 2	Large, yellow
		Pete 3	(doesn't know)
		Unguja Ukuu (group)	Large, tall and long, yellow, with largish black spots

Table 3.1 (continued)

Swahili name	Literal translation (if transparent)	Informant(s) (by residence)	Description given (salient features)
		Uzi 1	Larger and whiter than *chui uwanda*, with large patches
		Uzi 2	With large spots of every colour, the fiercest type
unyasi	'Grass (-like leopard)'	Muyuni	Small, long, with more yellow than *konga* and small spots
		Pete 2	Smaller and darker than *kisutu*, with no white
		Pete 3	Red, with small white spots
		Uzi 2	Reddish, like the colour of (dry) grass
chui unyasi	'Grass (-like) leopard'	Pete 1	Yellowish

Source: Walsh 1997; 2000; 2007: appendix, table A3.

to verify these general statements or compare them with local descriptions. Only three of the specimens are sexed: the type specimen in London is described as a young adult male, while the two Harvard specimens are females presumed to be full adults. The male type has a much longer tail and so total body length than the females, and also has a broader (if not longer) skull. It also has a slightly different coat pattern, but in the absence of a much larger sample, including other known males, as well as more females and individuals of various ages to compare it with, it is impossible to hypothesise whether this is because of its sex or age, or is an individual feature. Otherwise it is difficult to evaluate the colour of these specimens because of the different ways in which they may have been preserved. The limited available evidence suggests that Zanzibar leopards exhibit sexual dimorphism similar to other leopard populations in Africa and Asia, but there is little that we can currently add to this.

When asked, hunters denied that different named varieties of leopard corresponded to the different sexes. One knowledgeable informant in Kitogani opined that *kisutu* leopards (which he contrasted with the larger *konge* variety) were more likely to be female, and recalled a male *konge* that had been killed by the National Hunters (*Wasasi wa Kitaifa*) in 1989 near the main road to Paje. But an interviewee in Pete described a male *kisutu* that he had shot near Pete in 1979 with the National Hunters: in his classification the larger *kisutu* type was contrasted with the smaller *unyasi*. Others described individual *konga/konge* leopards as either male (an informant in Muyuni) or female (an informant in Makunduchi), while another man in Makunduchi remembered killing both male and female *kichigi* leopards during a long career as a hunter. As well as there being no general correlation between named variety and sex, hunters also offered different opinions on whether the different kinds could interbreed. And some, as we have already noted, denied that there were different varieties at all. Otherwise most informants who claimed to be able to distinguish between male and female leopards in the field did so on the basis of their behaviour (for example by their different vocalisations) rather than on their size and build.

Informants' identifications are difficult to test in the field because of the lack of both live animals and preserved whole skins. One of our research assistants, the former secretary of the National Hunters, had two small pieces of pelt taken from an unsexed leopard killed in Muyuni in 1986. He showed one of these pieces (illustrated in Walsh and Goldman 2008: 5, fig. 3) to some of our informants in the field. But whereas he identified this as being from a *mwanzi* leopard (distinguished by its colour as opposed to the large *mkonge* and the small *kichigi*), others classified it differently. A hunter in Jozani thought that it had come from a reddish-coloured variety of leopard that he did not know the name of, but which was different from the African civet-coloured *kisutu*. A hunter in Muungoni said that it was from a *konge* leopard (longer, and with a bigger head than the *kisutu*, which has more white on its coat), while a colleague in Kitogani used essentially the same name (*mikonge*, the plural of *mkonge*), a kind of leopard that he described as long and yellowish in colour, with some white on its coat (in contrast to a shorter variety, bluish in colour, that he could not name). The first and second identifications can be interpreted as having the same or similar referents, but this is clearly different from the *(m)konge* type described by the last two informants. This is the closest we have to a systematic experiment, and again the results suggest divergent views.

The different taxonomies of the Zanzibar leopard that we recorded cannot be reconciled into a single classification that correlates neatly with known differences between animals. Although not claiming to be cryptozoologists, Marshall and Selkow took a naïve cryptozoological approach to this ethnozoological problem by assuming that the names they were given mapped directly onto different varieties of leopard. As we have seen, they were mistaken, and the varieties they thought they were describing turn out to be imaginary.[9] Careful ethnozoological and linguistic research is required in cases like this (as exemplified by Forth's contribution to this volume). The cryptozoological question is still valid: what, if anything, do the categories in Zanzibari leopard taxonomies refer to? The status of the Zanzibar leopard and the state of knowledge about it is such that this has become a very difficult question to answer. The (past) existence of the leopard itself is not in doubt, but information on its morphology and the possible referents of local descriptions is hard to come by, let alone verify. As a result we may never fully understand these taxonomies, and may never shake off the suspicion that they refer to imaginary beings, or rather imaginary categories of the being that is (or was) the Zanzibar leopard.

Witchcraft and leopard-keeping narratives

Cutting across these individual taxonomies of leopard morphology is another, much more widely known, classification of leopards on Unguja. This is the distinction between 'wild' leopards (*chui wa mwituni*), and 'domesticated' leopards (*chui wa kufugwa*).[10] The latter are thought to be kept by witches (*wachawi, wabaya*) and used by them to intimidate their fellow villagers. When asked how they distinguish between wild and kept leopards, rural dwellers point out that wild leopards are only observed far away from human settlement and typically run away when disturbed. A leopard seen, heard or tracked in the vicinity of a village, or thought to have followed or attacked people or livestock, can be assumed to be a kept leopard. Sometimes, it is said, they are seen together with their owners, and then the status of both the leopard and its keeper is in no doubt. This is also the case when a leopard is seen that is wearing earrings, anklets, or other items of adornment and clothing. Informants suggest various proportions of wild to kept leopards, but agree that wild leopards can become kept leopards – when leopard keepers gain control over them through the use of magic – and kept leopards can become feral – when a leopard keeper dies without transferring his leopard to another keeper and the animal simply returns to the forest.

Belief in leopard-keeping provides a ready explanation for the appearance of leopards in settled areas and their occasional attacks upon livestock and

Table 3.2 The vocabulary of leopard-keeping narratives

Swahili term	Literal translation
Leopard (generic)	
chui [cʰui]	'Leopard'
Wild leopard	
chui mwitu	'Forest leopard'
chui wa mwitu	'Forest leopard'
Kept leopard	
chui wa fundi	'Specialist's leopard'
chui wa kufugwa	'Tamed, domesticated, kept leopard'
Leopard-keeping	
ufugaji chui	'Leopard-training, leopard-keeping'
Leopard keeper	
fundi	'Skilled person, specialist'
mfugaji	'Tamer, trainer, breeder'
mfuga chui	'Leopard trainer, breeder, keeper'
mwenye chui	'Leopard owner'
Witch (generic)	
mbaya	'Bad, evil person'
mchawi	'Witch, sorcerer'
Guild (generic)	
chama	'Club, association, guild' (including leopard keepers'/witches' guilds)

Source: 1996 fieldnotes, Goldman and Walsh 1997.

humans. This represents a specific and compelling development of more general ideas about witchcraft and sorcery in Zanzibar (Ingrams 1931: 465–477; Goldman 1996: 349–357; 371–378; Arnold 2003), ideas which themselves reflect patterns that are widespread in Sub-Saharan Africa and further afield (Evans-Pritchard 1937; Middleton and Winter 1963; Marwick 1970; Moore and Sanders 2001; Stewart and Strathern 2004; West 2005). Like witchcraft beliefs in general, they provide a handy explanation for particular cases of misfortune and other unusual events, can function as a tool for social maintenance and control, and possess a circular logic that is virtually unassailable in the context of local discourse – though individual doubters do exist. In the normal course of affairs alleged leopard keepers are not accused openly, but are the subjects of rumour, gossip and appropriate degrees of fear and respect. On occasion, though, leopard keepers and witches have been publicly accused in Zanzibar, giving these beliefs a more dynamic role in promoting social and political change. Some examples of this have already been mentioned, and are described in greater detail in an earlier paper (Goldman and Walsh 1997).

The standard narrative of leopard-keeping is elaborated in many ways, and incorporates details of how the leopards are obtained and bred, how they are kept and fed, and how they are trained and manipulated by their (mostly male) owners, who are thought to form leopard-sharing associations (*vyama*, singular *chama*) (Goldman and Walsh 1997: 5–15). Leopard keepers are said to obtain their leopards in a variety of ways: by inheriting them from kin, being given or sold them by a friend, relative, or more distant associate, or by habituating wild leopards. Wild leopards can be habituated by leaving out food to which magical ingredients (*dawa*) have been added. According to one of our informants, a novice acquires co-ownership of a leopard in the following way: enlisting the aid of an experienced leopard keeper, the would-be leopard owner goes with his friend out into the forest, where the keeper summons his leopard. The keeper's assurances overcome the leopard's reluctance to draw near a stranger, and over the course of several days the animal is habituated to the new person. Finally, the keeper commands the leopard to obey his new associate and the leopard waves its tail to show its consent. The novice is now the leopard's co-owner.

It is believed that a single leopard is commonly shared by several keepers, who may or may not be related and/or reside in the same community. For example, although it is said that there are no leopards on Tumbatu island, lying off the northwest shore of Unguja, it is also thought that some Tumbatu people may have 'shares' in leopards on the main island. The notion of leopard-keeping associations (*vyama*) is consistent with the more general Zanzibari belief (also held on Pemba island, where there are no leopards) that witches are highly social, if amoral, beings organised into cohesive guilds (*vyama*) with particular initiation processes, membership rules, headquarters (invisible to the uninitiated), hierarchies and group activities (Goldman 1996; Arnold 2003). Some of our informants asserted that leopard keepers control their leopards with the help of possessory spirits (*mashetani*), and it may be no coincidence that leopard-keeping itself is generally described in the same terms as the domestication and care of spirits by humans. Indeed, there is historical evidence suggesting that contemporary ideas about leopard-keeping derive from changing understandings of the role that village ritual specialists (*wavyale*, singular *mvyale*) once played as the guardians and propitiators of local spirits (*mizimu*), leopards, and other wild animals (for details see Walsh and Goldman 2007: 1141–1145). An old hunter in Charawe described being taken by his maternal grandfather's brother to a spirit shrine (*mzimu*) in a cave near the coast. They went there to make offerings and seek a cure for his wife's sickness. When they were there a leopard emerged from the cave: this wasn't really a leopard, he said, and it wasn't kept, but it was a *shetani* or spirit that would attack the witch that had caused his wife's illness. She subsequently recovered. Other interviewees told us that leopards could become invisible, like witches themselves.

Leopards are said to be kept by groups of witches, rather than individuals, as a way of protecting themselves from retribution. If a person believes that they have been the victim of witchcraft, including harassment by a kept leopard, they may announce that they will read a much-feared religious curse (*halbadiri*) that will condemn the perpetrator or leopard keeper to death (see Ingrams 1931: 472). In such cases, a single leopard owner will receive the full measure of the curse. However, a group of leopard keepers can shift their shared animal from owner to owner, and village to village, thereby dodging the effects of the curse. This belief can also be used to explain why, in some instances, no suspected witches in a community fall fatally ill immediately after the recitation of the curse.

Leopard keepers are believed to exercise control over their animals through the regular supply of magically doctored food to them. Several informants claimed to have spied leopard keepers provisioning leopards at some isolated spot in the bush. Kept leopards may reside in the bush, in which case they are summoned by their owners by knocking sticks or pieces of wood against one another or other objects. This activity is another which informants claim to have seen and, along with food provisioning, is commonly cited as evidence for the practice of leopard-keeping. Other leopards are thought to reside in their keepers' homes, usually in a special room and often under a bed. According to some informants, the sight of a leopard entering a home or disappearing in the immediate vicinity of a house is not uncommon and constitutes clear evidence of leopard-keeping. The translocation of kept leopards, as when an owner inhabiting one village 'lends' the leopard to a co-owner residing elsewhere, is effected through the transference of a small charm (*pingu*), sometimes unwittingly carried by a third party. Aside from the use of doctored food and such charms, the precise mechanisms through which kept leopards are supposed to be controlled and sent by their owners to do harm remain unelaborated.

Kept leopards are said to be deployed by their owners for a number of reasons: to terrorise people and induce respect; to guard livestock or other wealth; to obtain food in the form of meat; and to make money through breeding and the sale of their offspring. Harassment by kept leopards can take several forms, ranging from the animal simply allowing itself to be seen at a distance, to entering a village, mauling livestock, and savaging human beings. The object of harassment may be a single individual, a family, or an entire community. One informant explained that because an individual is embedded within a community, a leopard keeper cannot harass one person without frightening great numbers of others. Indeed, he continued, there have been cases of entire villages being abandoned as a result of particularly harrowing bouts of leopard harassment. It is widely believed that the mere sighting of a kept leopard causes the observer to become instantly ill, and that a particularly close encounter will result in leopard hairs appearing in the victim's vomit or faeces. The only remedy for afflictions of this kind are traditional means of curing which typically involve divination in order to determine the source of the evil. Divination is not always necessary, however: it is believed that after someone has seen a leopard, its keeper usually visits the victim the following day, greeting him effusively and enquiring solicitously about his health and welfare. Such conduct – which an outsider might consider part and parcel of normal Zanzibari social behaviour – is considered a flimsily disguised menace that serves to inform the victim that his outwardly amiable interlocutor was the person responsible for the leopard encounter and that it would be wise for the victim not to cross the keeper again in any way.

Due to their association with witchcraft and the threat they pose to life, limb, and livestock, leopards are hated and feared by rural Zanzibaris, including the hunters who encounter and sometimes kill them by accident. Though some stated that a leopard owner need only touch the body of his killed leopard and the hunter will rumble like a leopard or even die, others described a range of protective measures that could be taken, including the removal of certain body parts from a dead leopard. The larynx (*koromeo*) or throat (*koo*), for example, must be excised lest the owner discover the carcass and use the parts as the basis for a powerful hex on the leopard killer. In addition, a leopard killer sometimes removes other parts such as the claws that can be employed to concoct a magical recipe that will protect him from future leopard encounters. Such prophylactic measures are also applied to livestock and especially hunting dogs. Finally, it should be noted here that kept leopards are themselves believed to be protected by magic. A number of informants claimed that this magic causes hunters' guns to jam, though others said this was not so: the age and state of many guns currently employed in Zanzibar makes malfunctions quite likely. Nonetheless, the fear induced by leopards and belief in leopard-keeping are such that they resulted in determined attempts to eliminate leopards from Unguja island. As we have described in detail elsewhere (Walsh and Goldman 2007), localised hunting and trapping culminated in successive nationwide campaigns of extermination after the Zanzibar Revolution, when colonial controls upon leopard-killing were ignored.

When we began our joint research in 1996 we were asked, among other things, to examine the claim that the Zanzibar leopard was domesticated. Simplified leopard-keeping narratives were related so frequently and convincingly that a number of researchers believed them to be factual, or at least worth investigating further (see Walsh and Goldman 2012). However, as soon as we heard the detailed accounts that most other outsiders had not had access to, it was evident that they were too fantastic to be true, but were similar to other witchcraft beliefs and could be understood accordingly. From this (classic anthropological) point of view, both the witches and the kept leopards of Unguja are imaginary beings: unless under duress or suffering from delusion no one in Zanzibar will describe themselves as a witch or leopard-keeper, and there is no evidence at all for the existence of domesticated leopards and leopard-keeping, not to mention breeding and other practices believed to be undertaken by keepers' guilds.[11] Unfortunately (from a conservationist point of view) these beliefs have accelerated the demise of the Zanzibar leopard, and turned it into a more familiar kind of cryptid, an animal that some believe to be extinct but others continue to see. This is the subject we turn to next.

The perceived persistence of the Zanzibar leopard

When we undertook our consultancy in July 1996 there appeared to be good evidence for the continuing presence of the Zanzibar leopard in the south and east of Unguja island. National Hunters' reports indicated that leopards were still being killed (along with other 'vermin'): the last kill on record was from a hunt in Jambiani on 17–18 April 1995 (Goldman and Walsh 1997: 31–36). We were also told of more recent kills by independent groups of hunters. The former secretary of the National Hunters, who worked as our research assistant, told us that 'Omani Arab' hunters from Mlandege in Zanzibar town were rumoured to have killed a leopard at Mtule, between Kitogani and Paje, in March 1996 (Goldman and Walsh 1997: 29). An interviewee in Dimani told us that three leopard cubs had been killed by young hunters in that area on or around 21 April 1996 (Goldman and Walsh 1997: 26). These and reports of earlier kills were supplemented by descriptions of recent leopard sightings and other evidence for their presence, including seven reported sightings in 1996 (Goldman and Walsh 1997: 3, 24–36, 2002: 19–22).

At the time we had no reason to disbelieve these claims, though some experienced hunters said that they had not seen evidence of leopards for a number of years. We knew that most kills by independent and local groups of hunters would be hidden from official view and would not be included in National Hunters' statistics. And their former secretary – who showed us his own list of kills – told us that the National Hunters did not submit records of all of their leopard kills because they were concerned that the hunting of leopards might be stopped by the government. As it happens, when we began our joint research moves were afoot to do just this. The recently passed Forest Resources Management and Conservation Act of 1996 provided for the preparation of lists of protected wild animals and plants, and, in the interim, the use of lists that had been on the statutes since the colonial period but ignored since the Zanzibar Revolution (sections 76–79, part VIII, 'Conservation of Wild Animals and Wild Plants', in The Revolutionary Government of Zanzibar 1996: 40–42). The Zanzibar leopard was on the old schedule of protected wild animals (Walsh and Goldman 2007: 1140–1141, 1150), and was about to be included in the new one. As a result it had become illegal (again) to kill, injure, destroy, capture or collect leopards without a special permit, unless this was done 'to defend against an attack or imminent threat of attack on human life' (section 77 in The Revolutionary Government of Zanzibar 1996: 40).

This new legislation seems to have had an immediate effect on the practice of the National Hunters. From January 1996 onwards the vermin hunting statistics based on their reports no longer included leopard kills.[12] Since July 1996 we have not been told of any leopard kills by National Hunters. Nor have we heard of any certainly killed by local groups, individual hunters, or others. In August 1999 the then Head of Conservation in Zanzibar told Goldman that he had heard that a dead leopard had been found near Kinyasini about a year earlier (i.e. sometime in mid-to-late 1998), but he knew no more details, including whether this leopard had been killed or not. We do not know whether the lack of information about leopard kills in the past eighteen years reflects reluctance to talk about an activity that has been (re)defined as illegal, or the fact that very few and perhaps no leopards at all have been killed since 1996. It may be that there are very few or no leopards left to kill on Unguja. Otherwise we wonder whether we would have picked up on more stories of kills had we spent longer in the field or interviewed more hunters during subsequent research trips and other visits to Zanzibar.

At the same time, reports of leopard sightings and other evidence for their presence continue to reach us. Every time we visit Zanzibar and meet with former colleagues working in the Department of Commercial Crops, Fruits and Forestry (DCCFF) we are titillated with tales of sightings and other incidents, some of them involving kept leopards and their alleged keepers, others more prosaic, and so to us more believable. During a visit in October 2007, Walsh was told at the headquarters of Jozani–Chwaka Bay National Park that about three months earlier a ranger on night patrol had reported seeing a tree shaking and then a half-eaten Blue monkey (*Cercopithecus mitis*) – presumed to be a leopard's meal – lodged in its branches. The details of this, it was said, had been recorded by the Chief Park Warden (who was one of our research assistants in 1996), and in a subsequent interview the latter recounted that the ranger in question was a forty-year-old man from Ukongoroni, and that what he had actually seen was not a monkey, but a leopard in a fork of a tree about four metres off the ground. This was not at night, but around ten o'clock in the morning at a place called Kiwandani. The Chief Park Warden was not sure of the date: he had maybe interviewed the ranger in April, and perhaps the incident had taken place in January 2007. The ranger was very excited when he came to tell him what he had seen. He described making eye contact with the leopard, whereupon he began to step slowly backwards, thinking that the leopard's keeper might be somewhere near. He also related how he had watched the leopard stretching its way down the trunk of the tree after leaving the fork in which it was resting. He had not seen a leopard for many years, but claimed not to have been frightened by this unexpected encounter.

It is instructive here to note the discrepancies between the second- and third-hand reports of this sighting: the ranger involved would presumably have given a different account again. In the same interview the Chief Park Warden also mentioned that leopards had been much in evidence in Makunduchi in 2007: they were reported to have preyed on chickens, ducks and goats, and villagers had responded by reading the *halbadiri* curse (see above) against the leopard keepers, whoever they were. Other reports of leopard sightings have come to us by email. In February 2009 the Zanzibari owner of Zala Park, a small private zoo in Muungoni, wrote to say that a leopard had been heard at Mtule on the Kitogani–Paje road, and that the watchman at a brickmakers claimed to have seen it twice, in December 2008 and January 2009. In March 2009 we were copied into correspondence by the Chief Park Warden reporting that since the start of the year four rangers had independently seen a leopard at a single location in the forest in Jozani–Chwaka Bay National Park. The last sighting was on 1 March. We were unable, however, to obtain further details about these sightings. It is difficult to elicit information on cases like this at a distance, though our former assistant does make an effort to record particular incidents. Indeed he was also involved in the following case, which provides further illustration of the difficulty that we have in verifying reports and reconciling the accounts of different informants.

In April 2002 this officer told Walsh that in August 2001, while undertaking a survey of the mangrove forest northwest of Jozani, he, other staff, and the local villagers who were also members of the surveying team, had come across leopard tracks at Wangwani. They followed the spoor until they encountered the remains of a male Suni antelope (*Neotragus moschatus*), which they assumed to have been killed by the leopard. He collected one of its horns and also leopard faeces from the site, and took them back to Forestry Commission headquarters at Maruhubi, where they were kept in bottles. However, when Walsh asked to see this material, it could not be found, and he was led to understand that the bottles had probably been thrown away by cleaners or other staff, perhaps afraid of the presence of leopard-related relics in their workplace. Walsh followed this up by interviewing a Jozani forest guard who had been present when the finds had been made at Wangwani. He provided a detailed account that differed in a number of ways from the first one. He agreed, however, that they had collected leopard faeces, adding that the men on the team (eight in all) were quite apprehensive about this, afraid that it was a kept leopard and knowing that keeping such objects was a dangerous thing to do. The first officer (our former assistant) later said that he had kept this material on his desk for some time, but that it had indeed been thrown out by the office cleaners. It emerged that the same fate had also befallen presumed leopard scat collected by Goldman and handed over to the office in 1997.[13] In January 2003 Goldman followed up on the Wangwani case. Two of the other men who had been on the original survey team took her to the site of the antelope kill, and provided accounts that were at variance with both of those given to Walsh. They could not cast any light on the loss of the material taken to Maruhubi, but like others in the party were clearly steeped in leopard-keeping lore and not entirely happy with the collection of leopard faeces. Further efforts by Goldman to find out more information produced only more discrepancies, but did at least result in the finding of the original data record of the mangrove survey (Walsh and Goldman 2012: 736–737).

Despite the discrepancies, and the allusions to leopard-keeping, it is easy to imagine a core of truth in these accounts, and to hypothesise that they were based on real sightings and/or signs of a leopard's presence – or another animal mistaken for a leopard. But other claims stretch credibility. We have written at length elsewhere (Walsh and Goldman 2012) about the 'kept leopard chases' that a number of researchers and visitors to Zanzibar have taken part in and sometimes paid money for, lured by the promise of being shown a tame leopard. These quixotic quests for imaginary animals have always proved fruitless, much to the chagrin of their pursuers. In recent years Zanzibari conservationists have joined the pursuit along with non-Zanzibaris beguiled by leopard-keeping narratives. In January 2003 DCCFF staff told Goldman about an ongoing kept leopard chase that had involved a number of them. According to one official in the department, it began three months earlier with reports of leopard predation on livestock. Another official in the same department denied that there had been reports of leopard attacks on livestock. Instead he claimed that the case had been brought to the attention of the DCCFF because a leopard had been seen entering and leaving a house in Marumbi, and the person who had observed this had been bewitched and struck mute. DCCFF employees were sent to follow up on these reports, and one of them saw the house of the supposed leopard keeper together with a peculiar opening at the back that would allow a leopard to come and go. The visitors met with the deputy *Sheha* (local administrator) to talk about events, but were spooked into silence when they realised that the alleged leopard keeper was lurking nearby. A DCCFF team returned again in January 2003, and their leader asked the deputy *Sheha* to collaborate in an attempt to get a photo of the kept leopard. The deputy was very reluctant to agree to this and clearly afraid, but was told that as a government employee he had no option but to cooperate. Eventually he yielded, but it was agreed that when the researchers returned with a camera they would have to pretend that they were doing something else, like surveying monkeys or birds. Back in Zanzibar town, the team leader approached Goldman, who was then photo-trapping in Jozani forest, hoping that she would carry out the plan, for which transport for a team of DCCFF staff as well as money for accommodation and a payment to the *Sheha* were all required. While she considered the limited time and resources at her disposal – and the implications of the discrepancies in different accounts, together with a rumour that the leopard in question had been shifted to another location – the team leader promised to write a letter of introduction that would smooth the way with the local administrator. This letter was never delivered and there were other indications that the DCCFF had dropped their plan to follow up on this case. That was the last we heard of the Marumbi leopard.

Whereas foreign researchers have sometimes been beguiled into believing that leopard-keeping is really practised on Unguja, some Zanzibari researchers and others have sought to reconcile their own witchcraft beliefs with conservation science by proposing that kept leopards be displayed to the public and tourists in particular. The idea of a zoo or holding pen for leopards was suggested by one of the American student researchers referred to earlier, Benjamin Selkow (1995: 12); it has since been taken up enthusiastically by Zanzibaris believing that leopard keepers might make good use of their leopards in this way. One of our research assistants in 1996, the former secretary of the National Hunters, asked a number of our interviewees whether it would be feasible to persuade leopard keepers to display their leopards to the public and fee-paying tourists. The same idea also came up in the discussion that followed our end-of-fieldwork presentation to the Jozani–Chwaka Bay Conservation Project (JCBCP) and other government staff (Walsh 1996). We were careful in that meeting not to overtly criticise the beliefs of the many people in the audience who believed in leopard-keeping, and we did not question the display proposal as directly as we might have done otherwise. In October 1996 Goldman and a colleague in the Commission for Natural Resources were asked to investigate just such a proposition. They met with four men in Kizimkazi who wanted to capture and display leopards: they claimed to have seen a leopard in the area in recent months, and thought it possible that this and indeed all leopards were kept. But they had no clear idea of how they would obtain and care for the animals, other than making a range of suggestions (for details see Walsh and Goldman 2012: 738). The proposal did not receive official approval and Goldman (who was then working for JCBCP) heard nothing more of this scheme.

The leopard display idea clearly did not wither and die after the dissemination of our final report, in which we made it clear that we thought leopard-keeping to be wholly imaginary (Goldman and Walsh 1997: iii, 1, 13–15). Indeed, it surfaced in a quite unexpected place, in a debate in the Zanzibar House of Representatives in April 2003, when the Deputy Minister for Agriculture, Natural Resources, Environment and Cooperatives declared that his ministry would be happy to buy leopards to display them to tourists (Walsh and Goldman 2012: 739). As far as we are aware, no one has yet come forward to sell a live Zanzibar leopard to the government.[14] Needless to say, proposals for the display of leopards continue to be made by independent entrepreneurs such as the owner of Zala Park. In February 2009 he wrote to Goldman to tell her about his plan to build a leopard enclosure at a site in Pete. He asked for any advice she could give on keeping leopards for the purposes of conservation, education and research. In a subsequent communication he declared that he had been forced to put his plans for the new facility on hold because he had been denied permission by the authorities. Expansion of the existing zoo at Muungoni would also require official approval if he was to display animals protected under the Forest Resources Management and Conservation Act of 1996.

These contemporary notions about the possible social and economic benefits of displaying captive leopards are indicative of the continuing strength of the leopard-keeping narratives that underpin them. Meanwhile, we can hypothesise that the cultural salience of leopards on Unguja, the widespread belief in the existence of leopard-keeping, and the consequent apprehension that many people feel about this form of witchcraft, result in many more imagined sightings and claims regarding leopards than would otherwise be the case. Hunters themselves note that some of their colleagues and fellow villagers are liable to mistake other, smaller carnivores on the island for leopards.[15] These include the spotted and banded African civet (*Civettictis civetta*), which is explicitly compared to the Zanzibar leopard (see Table 3.1), and the Zanzibar servaline genet (*Genetta servalina archeri*), known in some parts of Unguja as *uchui* (Goldman et al. 2004: 6), literally 'the slender leopard'.[16] The tracks of different local carnivores (and sometimes other animals) can also be readily confused, for example when they have been made in sand or on dusty ground. During their 1997 survey, the Stuarts, who had authored a field guide to tracks and signs (1994), were disappointed by a number of such misidentifications. They cast doubt on the identity of the alleged leopard pugmarks illustrated in our printed report (Goldman and Walsh 1997: 56), and concluded that many islanders erroneously attribute the tracks of the African civet to the leopard (1997a: 4; 1997b: 1; 1998: 37).[17]

There is a large literature on the psychology of eyewitness testimony (for example Kapardis 2003: 21–125), and a smaller literature on the evaluation of claimed animal sightings, some of it written by cryptozoologists (for example Rabbit 2002). It is widely acknowledged that preconceptions can affect perceptions, and there is little doubt that this has happened in the case of the Zanzibar leopard, generating both false sightings and the misinterpretation of tracks and perhaps also other signs. Kept leopards and leopard keepers are imaginary beings, and belief in them has certainly contributed to imagined sightings, mistaken identifications, and the proliferation of gossip and rumour relating to these (compare Stewart and Strathern 2004). But some false attributions would probably be made anyway, for example when other animals or indirect evidence for their presence are confused with the leopard and its presumed signs. It is also possible that some misidentifications are not involuntary, but fabricated or otherwise elaborated by informants eager to please researchers, as the Stuarts argued (1997a: 4). We have seen examples of this ourselves when asking people to identify photographs of animals, a procedure which suffers from a number of pitfalls (see Barley 1983: 96–97; Diamond 1989).

This leads us to ask whether all recent reports of leopard sightings and signs can be explained away, as the Stuarts suggested, although even they allowed that a few individual leopards might survive (1997a: 3; 1997b: 1). As we have seen, some reports are more convincing than others. Leopards elsewhere are known to be largely nocturnal and secretive animals, capable of roaming through city suburbs as well as through the African countryside without being observed (Guggisberg 1975: 228; Hurn 2009: 6; Seidensticker 1991: 107; Nowell and Jackson 1996: 28; Sunquist and Sunquist 2002: 321). With this in mind, we cannot entirely rule out the possibility that the Zanzibar leopard does persist, as we stated in an earlier paper (Goldman and Walsh 2002). The same is suggested by the relatively recent discovery of two nocturnal carnivores on Unguja: the Zanzibar servaline genet in 1995 (Van Rompaey and Colyn 1998; Goldman and Winther-Hansen 2003a; 2003b), and a local population of the African palm civet (*Nandinia binotata*) in 1998–99 (Perkin 2004). Both of these were previously unrecorded, though they were known to villagers (Goldman et al. 2004).

Discussion and conclusion

The title of Jared Diamond's Extant unless proven extinct? Or, extinct unless proven extant? (1987) summed up a question that continues to be debated by conservation biologists (for example Brussard 1986; King 1988; Mace and Collar 1995; Reed 1996; Kéry 2002; Butchart et al. 2006; Roberts et al. 2009; Vogel et al. 2009). How is extinction in the recent past to be recognised or inferred? As Diamond and others have observed, detecting extinction is often easier said than done. The various qualitative and quantitative answers given to this question have influenced successive editions of the IUCN Red List of Threatened Species.[18] The *Guidelines for Using the IUCN Red List Categories and Criteria* (Version 8.1) state that:

> A taxon is Extinct when there is no reasonable doubt that the last individual has died. A taxon is presumed Extinct when exhaustive surveys in known and/or expected habitat, at appropriate times (diurnal, seasonal, annual), throughout its historic range have failed to record an individual. Surveys should be over a time frame appropriate to the taxon's life cycles and life form.
>
> (IUCN 2010: 9)

Operationalising this definition, however, and drawing a sharp dividing line between the IUCN categories of Critically Endangered and Extinct, has proved problematic. One problem is that premature declarations of demise can lead to what has been called the 'Romeo error', whereby 'any protective measures and funding are removed from threatened species in the mistaken belief that they are already extinct' (IUCN 2010: 67, citing Collar 1998). In response to discussions of this problem (see Butchart et al. 2006), the IUCN has introduced a qualifying tag so that taxa of indeterminate status can be described as 'Critically Endangered (possibly extinct)':

> Critically Endangered (possibly extinct) taxa are those that are, on the balance of evidence, likely to be extinct, but for which there is a small chance that they may be extant. Hence they should not be listed as Extinct until adequate surveys have failed to record the species and local or unconfirmed reports have been investigated and discounted.
>
> (IUCN 2010: 67)

If we had to choose a category (and tag) for the Zanzibar leopard it would probably be this one.[19] But we are also aware that a single corroborated sighting (for example supported by photographic evidence), or a physical specimen (and genetic profile) with well-documented provenance, would be sufficient to remove the 'possibly extinct' tag, at least in the short term.

Compare the Stuarts' conclusion following their 1997 survey:

> We encountered absolutely no sign of leopards during the survey and we believe that this cat is now extinct on the island, or at best present in such

low numbers that there is little, or no, hope of doing anything to save it in the wild state.

(Stuart and Stuart 1997b: 1)

Although they allowed for the survival of leopards, this pessimistic statement, as we have seen, discouraged further research and efforts to develop a leopard conservation programme.[20] The Stuarts' search for leopards and leopard sign, which included camera-trapping, together with a later photo-trapping survey conducted in Jozani forest by Goldman and Winther-Hansen (2003a), have been the only systematic attempts to detect the Zanzibar leopard to date. These and other camera-trapping exercises on the island do not constitute the 'exhaustive surveys in known and/or expected habitat' without any record of an individual that are required if a taxon is to satisfy IUCN criteria for extinction. The Stuarts were sceptical about the evidence for recent leopard kills and sightings that we recorded in our 1997 report, and discounted other reports that they heard in the field without thoroughly investigating these. Nor, it must be said, have we had the opportunity to follow up many of the accounts that we have been given since 1997 that might be counted as 'reasonably convincing recent local reports or unconfirmed sightings' and that would support the categorisation of the Zanzibar leopard as 'Critically Endangered (possibly extinct)' (IUCN 2010: 68). Needless to say, it would be difficult to apply probabilistic methods for inferring extinction (for example Roberts et al. 2009; Vogel et al. 2009) to these reports, given questions about their reliability, and the fact that many of them are not first-hand. Likewise any attempt to use the statistics of leopard kills for the same purpose would run into difficulties because they represent kills by only one group – the National Hunters – and because they appear not to have been kept once it became illegal again to kill leopards. Even if these methods were applied, it would only take a single corroborated observation, sample or specimen to falsify a hypothesis of extinction. Argument then shifts to determining what exactly counts as corroboration and so adequate evidence for falsification (see Roberts et al. 2009), and brings us back to assessing the reliability of Zanzibaris' claims to have seen leopards and other signs of their presence.

Cryptozoology flourishes in this epistemological abyss. It is interesting in this regard to compare the case of the Zanzibar leopard with that of the Tasmanian tiger, perhaps the best-known example of a carnivore considered extinct by zoologists but not by cryptozoologists. The last Tasmanian tiger (*Thylacinus cynocephalus*) in captivity died in Hobart Zoo in Tasmania on the night of 7 September 1936, by which time this species of marsupial carnivore may already have been extinct in the wild (Paddle 2000: 175–179, 195–197; Owen 2003: 124–134). It was declared officially extinct fifty years later, this being the period of grace then required by the IUCN Conservation Monitoring Centre (McKnight 2008). However, there has been no shortage of unverified sightings since 1936, both on Tasmania (for example Bailey 2001; Mittelbach and Crewdson 2005) and the Australian mainland (for example Smith 1996:

94–115), where there is evidence for the persistence of thylacines into the first half of the nineteenth century (Paddle 2000: 22–24):

> There are many well-meaning individuals, not just in Tasmania, but in all mainland states as well, who, to the present day return from driving or walking in wilderness areas with stories of thylacine sightings. But the criterion for establishing the existence of the thylacine beyond 1936 can only be met through the production of a body, either dead, or, preferably, alive.
>
> (Paddle 2000: 23)

This strict criterion derives from bitter experience. More than a thousand thylacine sightings have been logged with different authorities since 1936, but on further investigation they have dissolved into a melange of fantasy, fakes and false trails (Paddle 2000: 197; Owen 2003: 186–200). As in the case of the Zanzibar leopard, neither qualitative nor quantitative assessments are able to resolve these reports when hard physical evidence is lacking.[21] Likewise there would be no point in applying probabilistic methods of analysis to reported sightings of the Tasmanian tiger. One group of researchers has developed a bioeconomic model based on historical records of thylacine kills to re-evaluate the likelihood that the species has survived on Tasmania (Bulte et al. 2003). But their optimistic inferences do not falsify the hypothesis of thylacine extinction in the 1930s. They only make it less likely – and then only if the model's assumptions and parameters are deemed reasonable. This kind of modelling and projection is no substitute for hard evidence, though it has its attractions when hard evidence is hard to come by.

There are a number of similarities, as well as important differences, between the Zanzibar and Tasmanian cases. As is well known, European settlers on Tasmania had driven the indigenous human population to near extinction just as they set about exterminating the island's thylacines (Davies 1973; Ryan 1996). As a result we possess scant information on the ethnozoology of different Aboriginal Tasmanian groups, including their knowledge and practices relating to the Tasmanian tiger, though it seems that for some groups it had religious and cultural significance over and above its value as a possible item of food and whatever threat it posed to them as the largest predator on the island (Paddle 2000: 18–20). But we do know what the European community and in particular sheep farmers thought about thylacines. They were blamed – wrongly as it turns out – for killing large numbers of sheep, and widespread anger over stock losses led to the thylacine bounties and hunting that accelerated its demise. The Tasmanian tiger was almost universally loathed, and demonised as a ferocious blood-sucking predator (Paddle 2000: 29–35; Bagust 2003; Freeman 2007). These European representations of the thylacine parallel local narratives about the Zanzibar leopard and their alleged keepers, and it is no accident that they both resulted in island-wide campaigns of extermination. And it is very likely that the cultural salience and demonisation of both of these imaginary carnivores has, at different times, contributed to their perceived persistence and the sightings that seem(ed) to confirm this.[22]

The Tasmanian tiger was, of course, hunted and trapped to the brink of extinction in the early years of the twentieth century, and it has been presumed extinct for much longer than the Zanzibar leopard, whose status remains rather more equivocal. Relatively few people survive from the generation that last had direct experience of the thylacine, and the dominant narrative about it has undergone a remarkable turnaround in response to the development of mass tourism, the influence of environmentalism and contemporary ideas about the value of biodiversity, and, last but not least, the eager attentions of cryptozoologists (Owen 2003: 135–185; Mittelbach and Crewdson 2005; Turner 2009). Growing interest in the thylacine and the sea change in perceptions of it have had a dramatic impact on the number of recorded sightings, which increased exponentially in the second half of the twentieth century (Owen 2003: 187–189, citing Smith 1981). By contrast, the Zanzibar leopard is still widely feared by villagers, and it is too early to discern any trend in the number of alleged sightings. At the same time, it has already begun to attract the kind of interest from non-Zanzibaris that the Tasmanian tiger has long enjoyed. As we have seen, Zanzibari officials and conservationists have themselves proposed that leopards be kept and displayed in zoos. Leopard-keeping narratives are now highlighted in tourist guidebooks (for example McIntyre and McIntyre 2009: 54; Finke 2010: 136), and, as we have experienced ourselves, park staff and guides in Jozani–Chwaka Bay National Park are quick to tell stories about leopards and their owners to overseas visitors, and need no prompting to do so. Meanwhile, the Zanzibar leopard has also begun to feature on cryptozoological websites and internet forums, influenced in part by our own popular writing about it (for example Walsh and Goldman 2003, 2004).[23]

Cryptozoologists form a community with a special interest in speculation about the reality of particular kinds of imaginary being, including species that mainstream zoologists consider extinct. They are notoriously selective, focusing on salient creatures like the Tasmanian tiger but ignoring the vast majority of species that remain to be discovered and/or described by science. Cryptozoologists also typically only make selective use of the methods and literature of anthropology and the specialised discipline of ethnozoology. The case of the Zanzibar leopard illustrates some of the different ways in which a single animal can be imagined: through its imaginary taxonomy, its imagined domestication by other imaginary beings, and imaginary reports of its continuing existence. Anthropological discussions of animism and perspectivism provide other examples (for example Viveiros de Castro 1998; Bird-David 1999; Fausto 1999; 2004; Pedersen 2001). Cryptozoology, however, is largely concerned with simple ontological statements ('does it exist or not?'), and is generally lacking in epistemological and methodological sophistication, for example when it skips over the problems raised by the demand for falsifiability.[24] The claims of cryptozoology to respectability are further undermined by the ease by which some of its proponents extend its boundaries to include alien beings and other subjects that are usually classified as science fiction (for example Bord and Bord 1980). In these and other respects cryptozoology is naïve and justifiably described as a pseudoscience (see Simpson 1984). It would also comprise a fascinating subject for anthropological investigation.

We began this chapter with an epigraph taken from *The Book of Imaginary Beings* by Jorge Luis Borges and Margarita Guerrero. The passage we have quoted suggests that the universe of imaginary beings knows no bounds and is infinitely expandable (it is 'In brief, the sum of all things – the universe', 1969: 12), a wry observation that the authors quietly demonstrate in typical Borgesian fashion by imagining their own imaginary beings and providing these and other entries in their bestiary with the apparent, but often invented, apparatus of scholarship. This might also be adopted as the epigram for an anthropology of imaginary beings, in which the imaginary leopards and leopard keepers of Unguja take their place alongside other cryptids and their human (and not-so-human) analogues. Cryptozoology lacks the breadth of vision that this imaginary discipline would require. There is a nice example of this in one of the many cryptozoological compendiums compiled by Karl Shuker, in which he includes an entry for 'a grotesque sea monster termed the *hide*, documented by Jorge Luis Borges in his famous work *The Book of Imaginary Beings*' (2005: 79). Shuker's description of this legendary creature is taken from a short chapter on the 'Fauna of Chile' in the English translation of Borges and Guerrero's book (1969: 63–65). This was one of four new pieces written by Borges with his English translator, who has since made it clear that they are sprinkled with personal jokes (di Giovanni 2003: 27). Fortunately for Shuker, the paragraph about 'The *Hide*' (1969: 65) was translated word-for-word from a study of Chilean traditions that really exists (Vicuña Cifuentes 1910: 18). But he should not have relied on a work of fiction for his facts, at least not without tracking this story to its immediate source. This provides other accounts of the legendary '*El Cuero*', as well as information on their provenance – a good starting point for real anthropological and ethnozoological enquiry. We rest our case.

Acknowledgements

In addition to our numerous informants and all of the people we have thanked in earlier publications, we would like to extend our special gratitude to everyone else who has sent us information and stories about the Zanzibar leopard. Particular thanks are due to the following for their contributions to the research that has gone into the writing of this chapter: Mohammed Ayoub Haji, Peter Brussard, Anthony Cheke, Judith Chupasko, Greg Forth, Adrian Walsh, and all of our colleagues in the Department for Commercial Crops, Fruits and Forestry (DCCFF) in Zanzibar. The usual disclaimer applies.

Notes

1 The first edition of *The Book of Imaginary Beings*, published under the title *Manual de zoología fantástica*, was presumably largely written before Borges' breakup with Guerrero in 1952 (Williamson 2004: 319). The second edition, *El*

libro de los seres imaginarios (1967) was expanded and retitled, while the English translation by Norman Thomas di Giovanni (1969) was enlarged by Borges in collaboration with the translator himself (di Giovanni 2003: 27, 133). Heuvelmans' work first appeared in two volumes in 1955 as *Sur la piste des bêtes ignorées* (1955). Examples of cryptozoological 'bestiaries' inspired by the English translation and its subsequent revision (1962) and abridgement (1965) include Costello (1979), Shuker (1993), Coleman and Clark (1999), Eberhart (2002) and Arment (2004). Many similar works are listed in Shuker's bibliography (1999–), and a multimedia compilation would include large numbers of cryptozoological websites. The term 'cryptozoology' doesn't actually appear in *On the Track of Unknown Animals*, but was used by Heuvelmans in subsequent correspondence, following which he became known as the 'Father of Cryptozoology' (Coleman 2001; Eberhart 2002: xlvii).

2 This heterogeneous list is reminiscent of another, much discussed, Borgesian spoof, the classification of animals in 'a certain Chinese encyclopedia called the *Heavenly Emporium of Benevolent Knowledge*' (Borges 1999: 231; see Foucault 1970: xv–xxi; Lakoff 1987: 92–96). As we shall see, some Zanzibaris would assign the Zanzibar leopard to category (c) in Borges' imaginary classification, 'those that are trained', while sceptical observers might place it in (f) 'fabulous ones'.

3 For a more detailed review of our research on the Zanzibar leopard through to the present see Walsh and Goldman (2012: 729–730).

4 The historical overview in this and the following paragraphs draws on Walsh and Goldman (2007, 2012: 728–730) and the references cited therein.

5 'Within the period of the study I succeeded to see one wild leopard at Mapopwe study area [...] and Fatina Omari in 1981 (pers. comm.) sighted a leopard at Jozani study area' (Swai 1983: 53). Pakenham (1984: 49), citing an unpublished paper by the latter researcher, gives different dates: 'Leopards were observed as recently as 1980 and 1982 (Mturi 1983).' We haven't seen this paper to check whether they are referring to the same observations.

6 Note, however, that in reply to a question tabled in the Tanzanian parliament on 14 June 2007 about the possible reintroduction of animals that had disappeared from Zanzibar, Professor Jumanne Abdallah Magembe, Minister for Natural Resources and Tourism, stated that the Zanzibar leopard was 'the only animal which is believed to have disappeared from the forests of Zanzibar' (Bunge la Tanzania 2007: 13, our translation from the Swahili original).

7 Unguja island is generally agreed to have been separated from the African mainland since sea levels rose at the end of the last Ice Age; it is therefore presumed that the Zanzibar leopard has been genetically isolated from its continental relatives for the same period of time. The Sri Lankan leopard is thought to have been isolated for a similar length of time, whereas the Javan leopard displays much greater genetic divergence from other Asian subspecies (Miththapala et al. 1996: 1123–1124).

8 Only one of these interviewees was a woman, the informant in Ndudu who named one variety after the African civet (*ngawa*), and described the colour of another (*mwanzi*) by pointing to a nearby *kisutu* cloth. It may be no coincidence that *kisutu* is also the most widely known varietal name (see Table 3.1).

9 Given that they were students with limited training in zoology and anthropology, we do not mean to denigrate their work, which provided a valuable starting point for our own research. It might be noted, though, that they were both advised by Tony Archer, a wildlife consultant with much more zoological (if not ethnozoological) expertise (see Walsh and Goldman 2012).

10 This section draws on our original report (Goldman and Walsh 1997) and a paper on the history of leopard-killing (Walsh and Goldman 2007).

11 We are aware that leopards, like other wild animals, can be habituated to humans, particularly if they are hand-reared from a young age, but there is no documented

precedent, or rational explanation, for the behaviours described in leopard-keeping narratives. Moreover, when researchers have been offered the opportunity to see kept Zanzibar leopards (in exchange for a payment), the leopards have without exception failed to materialise. There is therefore not even hard evidence for the practice of keeping Zanzibar leopards as semi-domesticated pets, let alone for the other extraordinary claims that are made about kept leopards.

12 In addition to the statistics for the years 1983–95 analysed in our earlier work (Goldman and Walsh 1997; 2002), we have since obtained a complete record for 1995, and for the years 1996–99 and 2001–02.

13 Our interest in locating this material was prompted by the possibility that it might be used in genetic analysis, but nothing kept in the Maruhubi office appears to have survived, including the photographs of the leopard pelt and presumed pugmarks that were included in the original printed version of our 1997 report (Goldman and Walsh 1997: 55–56, figs 7–12): these particular images are therefore absent from the more widely distributed pdf version of the report.

14 Since we wrote this paper, however, at least one attempt has been made to pass off other live cats as specimens of the Zanzibar leopard. In September 2012 two young felids purported to be Zanzibar leopard cubs were brought into DCCFF offices and offered for sale for the princely sum of Tshs. 50 million each. It was immediately apparent that these were not leopard cubs and the would-be sellers were sent away. Subsequent examination of photographs of these animals suggested that they were the kittens of domestic cats.

15 On the other hand, there are a number of reports of a leopard being accidentally shot by a hunter who has caught no more than a fleeting glimpse of it in the dark, and assumed that it was a duiker or other animal.

16 The local dialect name *uchui* (earlier **luchui*) is derived from a combination of the Swahili class 11 noun prefix *u-* (**lu*), which typically signifies length and/or thinness, and the noun root *-chui*, 'leopard' (compare Nurse and Hinnebusch 1993: 349–351, 639).

17 Ordinarily, pawprints left by African civets and leopards are readily distinguishable because the former include claw marks whereas the latter do not: leopards keep their claws retracted when walking (Stuart and Stuart 1994: 17, 19, 26, 40). The Stuarts argued that the clawless pugmarks that are sometimes observed on paths in the forest and bush are signs of large African civets whose claws have been worn down by the rocky outcroppings prevalent on the coral rag of Unguja. Zanzibaris who are knowledgeable about wildlife deny this; according to them, African civet claws are never abraded to the extent that they leave no trace in pugmarks.

18 For the IUCN Red List and its background see www.iucnredlist.org/ (accessed: 30 April 2011).

19 Our Zanzibari informants, however, have offered a wide variety of opinions that would translate into the full range of IUCN categories and tags: Not Evaluated, Data Deficient, Least Concern, Near Threatened, Vulnerable, Endangered, Critically Endangered, Critically Endangered (possibly extinct in the wild), Critically Endangered (possibly extinct), Extinct in the Wild, and Extinct (IUCN 2010: 7–9, 67–70)!

20 This example suggests that applying the 'possibly extinct' tag might not always avert the Romeo error: declaring an animal extinct or almost certain to become so ('Going or gone' in the words of the title of Butchart et al. 2006) can kill off hope and funding for its conservation as effectively as pronouncing its definite extirpation.

21 As we have already indicated, we would also accept as evidence well-provenanced recent samples of faeces, hair, and other material that can provide DNA for genetic analysis. Likewise, clear photographs taken by researchers with professional

credentials and/or in the course of properly documented photo-trapping surveys are generally accepted as reliable evidence for an animal's presence.

22 Gregory Forth has suggested to us the interesting idea that extinct animals leave a 'shadow' in folk memory, such that the image of it is retained in discourse, influencing the interpretation of sightings and other signs by people who may be unaware that the animal in question is in fact extinct (personal communication, 12 March 2011). The names of animals do not necessarily follow them into oblivion, and will always continue in use for a time after their referents have disappeared. The same applies to lore about extinct creatures, and the cases of the Zanzibar leopard and Tasmanian tiger suggest that the persistence of such lore and its effects (which include false sightings) is at least in part a function of an animal's cultural salience. To the extent that the cultural significance of an animal does not depend on scientific or equivalent assessments of its status, then imagined sightings, rumour and gossip provide all the confirmation that is required to ensure that an extinct animal lives on in local discourse.

23 See, for example, the post 'Zanzibar leopard Do you think it still exists or now extinct' [sic], online at www.unexplained-mysteries.com/forum/index.php?showtop ic=193518 (accessed: 1 May 2010). Most of the text of this post is taken from the *Wikipedia* article on the Zanzibar leopard, http://en.wikipedia.org/wiki/Zanzibar_ Leopard, which cites our own publications.

24 For a sustained, but not entirely convincing, attempt to counter these and other charges against cryptozoology see the first part, entitled 'Science', of Chad Arment's *Cryptozoology: Science and Speculation* (2004: 9–156). The second part, headed 'Speculation', undermines the first (2004: 157–338).

References

Allen, G.M. 1939. A checklist of African mammals. *Bulletin of the Museum of Comparative Zoology at Harvard College*, 83, 1–763.

Ame, M. 2003. Serikali iko tayari kununua chui. *Zanzibar Leo*, 13 April, 6.

Anonymous. 1997. No sign of Zanzibar leopard. *Cat News* (Newsletter of the IUCN/ SSC Cat Specialist Group), 27, 12.

Archer, A.L. 1994. *A Survey of Hunting Techniques and the Results thereof on Two Species of Duiker and the Suni Antelopes in Zanzibar*. Report to FINNIDA/Forestry Sector, Commission for Natural Resources, Zanzibar.

Archer, A.L., Collins, S. and Brampton, I. 1991. Report on a visit to Jozani Forest, Zanzibar. *East Africa Natural History Society Bulletin*, 21(4), 59–66.

Arment, C. 2004. *Cryptozoology: Science and Speculation*. Lancaster, PA: Coachwhip Publications.

Arnold, N. 2003. Wazee wakijua mambo / Elders used to know things!: Occult powers and revolutionary history in Pemba, Zanzibar. Unpublished Ph.D. dissertation, Indiana University.

Bagust, P. 2003. Vampire dogs and marsupial hyenas: fear, myth and the Tasmanian tiger's extinction. In *Vampires: Myths and Metaphors of Enduring Evil*, edited by C.T. Kungl. Oxford: Inter-Disciplinary Press, 45–51.

Bailey, C. 2001. *Tiger Tales: Stories of the Tasmanian Tiger*. Sydney: HarperCollins.

Barley, N. 1983. *The Innocent Anthropologist: Notes from a Mud Hut*. London: Penguin.

Bird-David, N. 1999. 'Animism' revisited: personhood, environment, and relational epistemology. *Current Anthropology*, 40, S67–S91.

Bord, J. and Bord, C. 1980. *Alien Animals: A Worldwide Investigation*. London: Granada.

Borges, J.L. 1999 [1942]. John Wilkins' analytical language. In *Selected Non-fictions by Jorge Luis Borges*, edited by E. Weinberger. New York: Viking Penguin, 229–232.

Borges, J.L. and Guerrero, M. 1957. *Manual de zoología fantástica* (1st edn). Mexico City: Fondo de Cultura Ecónomica.

Borges, J.L. and Guerrero, M. 1967. *El libro de los seres imaginarios* (2nd edn, enlarged and retitled). Buenos Aires: Kier.

Borges, J.L. and Guerrero, M. 1969. *The Book of Imaginary Beings* ('Revised, enlarged and translated by Norman Thomas di Giovanni in collaboration with the author'). New York: E.P. Dutton.

Brussard, P.F. 1986. The likelihood of persistence of small populations of large animals and its implications for cryptozoology. *Cryptozoology*, 5, 38–46.

Bulte, E.H., Horan, R.D. and Shogren, J.F. 2003. Is the Tasmanian tiger extinct? A biological–economic re-evaluation. *Ecological Economics*, 45, 271–279.

Bunge la Tanzania 2007. [Tanzania Hansard], Majadiliano ya Bunge, Mkutano wa Nane, Kikao cha Tatu – Tarehe 14 Juni, 2007, Maswali na Majibu, Na. 25, Kutoweka kwa Baadhi ya Wanyama – Tanzania, 12–13.

Butchart, S.H.M., Stattersfield, A.J. and Brooks, T.M. 2006. Going or gone: defining 'possibly extinct' species to give a truer picture of recent extinctions. *Bulletin of the British Ornithologists' Club*, 126A, 7–24.

Chum, H. 1962. A vocabulary of the Kikae (Kimakunduchi, Kihadimu) dialect: with examples illustrating the morphology (edited by H.E. Lambert). *Swahili*, 33(1), 51–68.

Chum, H. 1994. *Msamiati wa Pekee wa Kikae: Kae Specific Vocabulary*. Uppsala, Sweden: Nordic Association of African Studies.

Coleman, L. 2001. Bernard Heuvelmans (1916–2001). Online. Available at: www.lor encoleman.com/bernard_heuvelmans_obituary.html (accessed: 21 March 2011).

Coleman, L. 2002. Introduction: if we don't search, we shall never discover. In *Mysterious Creatures: A Guide to Cryptozoology*, edited by G.M. Eberhart. Santa Barbara, CA: ABC-CLIO, xxxi–xxxiii.

Coleman, L. and Clark, J. 1999. *Cryptozoology A to Z: The Encyclopedia of Loch Monsters, Sasquatch, Chupacabras, and Other Authentic Mysteries of Nature*. New York: Fireside.

Collar, N.J. 1998. Extinction by assumption; or, the Romeo error on Cebu. *Oryx*, 32(4), 239–244.

Costello, P. 1979. *The Magic Zoo: The Natural History of Fabulous Animals*. London: Sphere Books.

Davies, D. 1973. *The Last of the Tasmanians*. London: Frederick Muller.

Dendle, P. 2006. Cryptozoology in the medieval and modern worlds. *Folklore*, 117, 190–206.

Diamond, J.M. 1987. Extant unless proven extinct? Or, extinct unless proven extant? *Conservation Biology*, 1(1), 77–79.

Diamond, J.M. 1989. The ethnobiologist's dilemma. *Natural History*, 6, 26–30.

di Giovanni, N.T. 2003. *The Lesson of the Master: On Borges and His Work*. London and New York: Continuum.

Dobroruka, L.J. 1964. Zur Verbreitung des 'Sansibar-Leoparden', *Panthera pardus adersi* Pocock.' Zeitschrift für Säugetierkunde 30, 144–146.

Eberhart, G.M. (ed.) 2002. *Mysterious Creatures: A Guide to Cryptozoology*. Santa Barbara, CA: ABC-CLIO.

Else, D. and Tyrrell, H. 2003. *Zanzibar: The Bradt Travel Guide* (5th edn). Chalfont St Peter, Buckinghamshire: Bradt Travel Guides.

Evans-Pritchard, E.E. 1937. *Witchcraft, Oracles and Magic among the Azande.* Oxford: Clarendon Press.

Fausto, C. 1999. Of enemies and pets: warfare and shamanism in Amazonia. *American Ethnologist*, 26(4), 933–956.

Fausto, C. 2004. A blend of blood and tobacco: shamans and jaguars among the Parakanã of Eastern Amazonia. In *In Darkness and Secrecy: The Anthropology of Assault Sorcery and Witchcraft in Amazonia*, edited by R. Wright. Durham NC and London: Duke University Press, 157–178.

Finke, J. 2010. *The Rough Guide to Zanzibar* (3rd edn). London: Rough Guides.

Forth, G. 2016. Cryptids, classification, and categories of cats: an ethnozoological study of unidentified felids from eastern Indonesia. In *Anthropology and Cryptozoology: Exploring Encounters with Mysterious Creatures*, edited by S. Hurn. Abingdon: Routledge.

Foucault, M. 1970 [1966]. *The Order of Things: An Archaeology of the Human Sciences.* London: Tavistock Publications.

Freeman, C. 2007. Imagining extinction: disclosure and revision in photographs of the thylacine (Tasmanian tiger). *Society and Animals*, 15, 241–256.

Goldman, H.V. 1996. A comparative study of Swahili in two rural communities in Pemba, Zanzibar, Tanzania. Unpublished Ph.D. dissertation, New York University.

Goldman, H.V. and Walsh, M.T. 1997. A leopard in jeopardy: an anthropological survey of practices and beliefs which threaten the survival of the Zanzibar leopard (*Panthera pardus adersi*). Zanzibar Forestry Technical Paper no. 63. Jozani–Chwaka Bay Conservation Project, Commission for Natural Resources, Zanzibar.

Goldman, H.V. and Walsh, M.T. 2002. Is the Zanzibar leopard (*Panthera pardus adersi*) extinct? *Journal of East African Natural History*, 91(1/2), 15–25. Map printed in the 2003 issue: *JEANH*, 92(1/2), 4.

Goldman, H.V., Walsh, M.T. and Winther-Hansen, J. 2004. Zanzibar's recently discovered servaline genet. *Nature East Africa*, 34, 5–7.

Goldman, H.V. and Winther-Hansen, J. 2003a. *The Small Carnivores of Unguja: Results of a Photo-Trapping Survey in Jozani Forest Reserve, Zanzibar, Tanzania.* Tromsø, Norway: privately printed.

Goldman, H.V. and Winther-Hansen, J. 2003b. First photographs of the Zanzibar servaline genet (*Genetta servalina archeri*) and other endemic subspecies on the island of Unguja, Tanzania. *Small Carnivore Conservation*, 29, 1–4.

Greenwell, R.J. 1985. A classificatory system for cryptozoology. *Cryptozoology*, 4, 1–14.

Guggisberg, C.A.W. 1975. *Wild Cats of the World.* Newton Abbot: David and Charles.

Harries, L. 1959. Supplementary vocabulary: Swahili–English. *Swahili*, 29(1), 55–80.

Hes, L. 1991. *The Leopards of Londolozi.* Cape Town, South Africa: Struik Winchester.

Heuvelmans, B. 1955. *Sur la piste des bêtes ignorées.* Paris: Librarie Plon.

Heuvelmans, B. 1958. *On the Track of Unknown Animals* (translated by R. Garnett). London: Rupert Hart-Davis. 2nd edn 1962, abridged edn 1965 (New York: Hill and Wang.)

Hurn, S. 2009. Here be dragons? No, big cats! Predator symbolism in rural West Wales. *Anthropology Today*, 25(1), 6–11.

Ingrams, W.H. 1931. *Zanzibar: Its History and Its People*. London: Frank Cass.

IUCN (International Union for Conservation of Nature and Natural Resources) 2010. Guidelines for using the IUCN Red List Categories and Criteria: Version 8.1 (August 2010). Prepared by the IUCN Standards and Petitions Subcommittee. Online. Available at: www.iucnredlist.org/ (accessed: 30 April 2011).

Jackson, P. 2000. Editorial: subspecies and conservation. *Cat News* (Newsletter of the IUCN/SSC Cat Specialist Group), 32, 1.

Johnson, F. (ed.) 1939. *A Standard Swahili–English Dictionary*. London: Oxford University Press.

Kapardis, A. 2003. *Psychology and Law: A Critical Introduction* (2nd edn). Cambridge: Cambridge University Press.

Kéry, M. 2002. Inferring the absence of a species: a case study of snakes. *Journal of Wildlife Management*, 66(2), 330–338.

Khamis, K.A. 1995. *Report on the Status of Zanzibar Leopards from 15th Dec. 1994 to June 1995 in Different Times at Zanzibar*. Certificate student paper, College of African Wildlife Management, Mweka, Tanzania.

King, F.W. 1988. Extant unless proven extinct: the international legal precedent. *Conservation Biology*, 2(4), 395–397.

Kingdon, J. 1977. *East African Mammals: An Atlas of Evolution in Africa. Vol. IIIA, Carnivores*. Chicago: University of Chicago Press.

Kingdon, J. 1989. *Island Africa: The Evolution of Africa's Rare Animals and Plants*. Princeton, NJ: Princeton University Press.

Kingdon, J. 2004. *The Kingdon Pocket Guide to African Mammals*. London: Christopher Helm.

Lakoff, G. 1987. *Women, Fire, and Dangerous Things: What Categories Reveal about the Mind*. Chicago: University of Chicago Press.

Mace, G.M. and Collar, N.J. 1995. Extinction risk assessment for birds through quantitative criteria. *Ibis*, 137, S240–246.

McIntyre, C. and McIntyre, S. 2009. *Zanzibar, Pemba, Mafia: The Bradt Travel Guide* (7th edn). Chalfont St Peter, Buckinghamshire: Bradt Travel Guides.

McKnight, M. 2008. *Thylacinus cynocephalus*. IUCN Red List. Online. Available at: www.iucnredlist.org/ (accessed: 1 May 2011).

Mansfield-Aders, W. 1920. The natural history of Zanzibar and Pemba. In *Zanzibar: The Island Metropolis of Eastern Africa*, edited by F.B. Pearce. London: Frank Cass, 326–339.

Marshall, S. 1994. The status of the Zanzibar leopard. Student paper, SIT Study Abroad, Zanzibar.

Marwick, M. (ed.) 1970. *Witchcraft and Sorcery: Selected Readings*. Harmondsworth: Penguin.

Meurger, M. 1988. Introduction. On the Lindorm's trail: the naturalization of culture. In *Lake Monster Traditions: A Cross-cultural Analysis*, by M. Meurger with C. Cagnon, London: Fortean Times, 9–36.

Middleton, J. and Winter, E.H. 1963. *Witchcraft and Sorcery in East Africa*. London: Routledge and Kegan Paul.

Miththapala, S., Seidensticker, J. and O'Brien, S.J. 1996. Phylogeographic subspecies recognition in leopards (*Panthera pardus*): molecular genetic variation. *Conservation Biology*, 10(4), 1115–1132.

Mittelbach, M. and Crewdson, M. 2005. *Carnivorous Nights: On the Trail of the Tasmanian Tiger*. Edinburgh: Canongate Books.

Moore, H.L. and Sanders, T. (eds) 2001. *Magical Interpretations, Material Realities: Modernity, Witchcraft and the Occult in Postcolonial Africa.* London: Routledge.

Moreau, R.E. and Pakenham, R.H.W. 1941. The land vertebrates of Pemba, Zanzibar and Mafia: a zoo-geographical study. *Proceedings of the Zoological Society of London* (Series A), 110, 97–128.

Mturi, F.A. 1983. Man's utilization of natural resources and the endangered Zanzibar red colobus monkey, *Colobus badius kirkii.* Unpublished paper, Department of Zoology, University of Dar es Salaam.

Mugo, K. (ed.) 2006. *Eastern Africa Coastal Forests Ecoregion: Strategic Framework for Conservation 2005–2025.* Nairobi: WWF Eastern Africa Regional Programme Office.

Nahonyo, C.L., Mwasumbi, L.B., Eliapenda, S., Msuya, C., Mwansasu, C., Suya, T.M., Mponda, B.O. and Kihaule, P. 2002. *Biodiversity Inventory of Jozani–Chwaka Proposed National Park, Zanzibar.* Technical report for CARE Tanzania / Department of Commercial Crops, Fruits and Forestry, Zanzibar.

Nowell, K. and Jackson, P. (eds) 1996. *Wild Cats: Status Survey and Conservation Action Plan.* Gland, Switzerland: IUCN.

Nurse, D. and Hinnebusch, T.J. 1993. *Swahili and Sabaki: A Linguistic History.* Berkeley: University of California Press.

O'Brien, S.J. and Johnson, W.E. 2005. Big cat genomics. *Annual Review of Genomics and Human Genetics,* 6, 407–429.

Owen, D. 2003. *Thylacine: The Tragic Tale of the Tasmanian Tiger.* Sydney: Allen and Unwin.

Paddle, R. 2000. *The Last Tasmanian Tiger: The History and Extinction of the Thylacine.* Cambridge: Cambridge University Press.

Pakenham, R.H.W. 1984. *The Mammals of Zanzibar and Pemba Islands.* Harpenden, Hertfordshire: privately printed.

Pedersen, M.A. 2001. Totemism, animism and North Asian indigenous ontologies. *Journal of the Royal Anthropological Institute* (N.S.), 7, 411–427.

Perkin, A. 2004. A new range record for the African palm civet *Nandinia binotata* (*Carnivora, Viverridae*) from Unguja Island, Zanzibar. *African Journal of Ecology,* 42, 232–234.

Pocock, R.I. 1932. The leopards of Africa. *Proceedings of the Zoological Society of London,* II, 543–591.

Rabbit, J. 2002. Native and western eyewitness testimony in cryptozoology. In *Mysterious Creatures: A Guide to Cryptozoology,* edited by G.M. Eberhart. Santa Barbara, CA: ABC-CLIO, xxxv–xliii.

Reed, J.M. 1996. Using statistical probability to increase confidence of inferring species extinction. *Conservation Biology,* 10(4), 1283–1285.

Roberts, D.L., Elphick, C.S. and Reed, J.M. 2009. Identifying anomalous reports of putatively extinct species and why it matters. *Conservation Biology,* 24(1), 189–196.

Ryan, L. 1996. *The Aboriginal Tasmanians* (2nd edn). St Leonards, NSW: Allen and Unwin.

Seidensticker, J. 1991. Leopards. In *Great Cats: Majestic Creatures of the Wild,* edited by J. Seidensticker and S. Lumpkin. Emmaus, PA: Rodale Press, 107–114.

Selkow, B. 1995. A survey of villager perceptions of the Zanzibar leopard. Student paper, SIT Study Abroad, Zanzibar.

Shuker, K. 1993. *The Lost Ark: New and Rediscovered Animals of the Twentieth Century.* London: HarperCollins.

Shuker, K. 1999–. *A Bibliography of Cryptozoological and Zoomythological Books.* Online. Available at: www.karlshuker.com/bibliography.htm (accessed: 21 March 2011).

Shuker, K. 2005. *From Flying Toads to Snakes with Wings.* London: Bounty Books.

Simpson, G.G. 1984. Mammals and cryptozoology. *Proceedings of the American Philosophical Society,* 128(1), 1–19.

Smith, M. 1996. *Bunyips and Bigfoots: In Search of Australia's Mystery Animals.* Alexandria, NSW: Millennium Books.

Smith, S.J. 1981. *The Tasmanian Tiger – 1980: A Report on an Investigation of the Current Status of the Thylacine* Thylacinus cynocephalus, *Funded by the World Wildlife Fund.* Unpublished report, National Parks and Wildlife Service, Hobart, Australia.

Smithers, R.H.N. 1971. Family Felidae. In *The Mammals of Africa: An Identification Manual,* edited by J.A.J. Meester and H.W. Setzer. Washington, DC: Smithsonian Institution Press, Part 8.1, 1–10.

Stewart, P.J. and Strathern, A. 2004. *Witchcraft, Sorcery, Rumors, and Gossip.* Cambridge: Cambridge University Press.

Stuart, C. and Stuart, T. 1994. *A Field Guide to the Tracks and Signs of Southern and East African Wildlife.* Cape Town: Southern Book Publishers.

Stuart, C. and Stuart, T. 1997a. *A Preliminary Faunal Survey of South-eastern Unguja (Zanzibar) with Special Emphasis on the Leopard* Panthera pardus adersi. Unpublished report, African–Arabian Wildlife Reserve Centre, Loxton, South Africa.

Stuart, C. and Stuart, T. 1997b. Zanzibar leopard: myth or reality? *The Arc* (newsletter of the African–Arabian Wildlife Reserve Centre), 3, 1–2.

Stuart, C. and Stuart, T. 1998. Unguja, Zanzibar's island of mystery. *Africa – Environment and Wildlife,* 6(5), 32–38. (with addendum in vol. 7(3), 11).

Sunquist, M.E. and Sunquist, F. 2002. *Wild Cats of the World.* Chicago: University of Chicago Press.

Swai, I.S. 1983. Wildlife conservation status in Zanzibar. Unpublished M.Sc. dissertation, University of Dar es Salaam.

Swynnerton, G.H. and Hayman, R.W. 1951. A check-list of the land mammals of Tanganyika Territory and Zanzibar Protectorate. *Journal of the East Africa Natural History Society,* 6–7, 274–392.

The Revolutionary Government of Zanzibar. 1996. The Forest Resources Management and Conservation Act, 1966. Zanzibar: Government Printer.

Turner, S.S. 2009. Negotiating nostalgia: the rhetoricity of thylacine representation in Tasmanian tourism. *Society and Animals,* 17, 97–114.

Uphyrkina, O., Johnson, W.E., Quigley, H., Miquelle, D., Marker, L., Bushs, M. and O'Brien, S.J. 2001. Phylogenetics, genome diversity and origin of modern leopard, *Panthera pardus. Molecular Ecology,* 10(11), 2617–2633.

Van Rompaey, H. and Colyn, M. 1998. A new servaline genet (*Carnivora, Viverridae*) from Zanzibar island. *South African Journal of Zoology,* 33(1), 42–46.

Vicuña Cifuentes, J. 1910. *Estudios de folk-lore chileno: Mitos y supersticiones recogidos de la tradición oral chilena, con referencias comparativas á los de otros países latinos.* Santiago de Chile: Imprenta Universitaria.

Viveiros de Castro, E. 1998. Cosmological deixis and Amerindian perspectivism. *Journal of the Royal Anthropological Institute* (N.S.), 4, 469–488.

Vogel, R.M., Hosking, J.R.M., Elphick, C.S., Roberts, D.L. and Reed, J.M. 2009. Goodness of fit of probability distributions for sightings as species approach extinction. *Bulletin of Mathematical Biology,* 71(3), 701–719.

Walsh, M.T. 1996. *The Zanzibar Leopard: An Anthropological Survey. End of Fieldwork Summary.* Report to Jozani–Chwaka Bay Conservation Project, CARE Tanzania, and Commission for Natural Resources, Zanzibar.

Walsh, M.T. 1997. Leopard linguistics and classification. Appendix C in H.V. Goldman and M.T. Walsh, A leopard in jeopardy: an anthropological survey of practices and beliefs which threaten the survival of the Zanzibar leopard (*Panthera pardus adersi*). Zanzibar Forestry Technical Paper no. 63. Jozani–Chwaka Bay Conservation Project, Commission for Natural Resources, Zanzibar, 37–52.

Walsh, M.T. 2000. Why does the Zanzibar leopard have so many names? Lexical variation and its interpretation on Unguja Island. Paper presented to the International Colloquium on Kiswahili in 2000, Institute of Kiswahili Research, University of Dar es Salaam, 20–23 March.

Walsh, M.T. 2007. Island subsistence: hunting, trapping and the translocation of wildlife in the western Indian Ocean. *Azania*, 42, 83–113, with online appendix at: www.biea.ac.uk/publications_pages/Walsh_appendix.pdf (accessed 3 May 2011).

Walsh, M.T. and Goldman, H.V. 2003. The Zanzibar leopard: between science and cryptozoology. *Nature East Africa*, 33(1/2), 14–16.

Walsh, M.T. and Goldman, H.V. 2004. The Zanzibar leopard: dead or alive? *Tanzanian Affairs*, 77, 20–23.

Walsh, M.T. and Goldman, H.V. 2007. Killing the king: the demonization and extermination of the Zanzibar leopard / Tuer le roi: la diabolisation et l'extermination du leopard de Zanzibar. In *Le symbolisme des animaux: l'animal clef-de-voûte dans la tradition orale et les interactions homme–nature / Animal Symbolism: The 'Keystone' Animal in Oral Tradition and Interactions between Humans and Nature*, edited by E. Dounias, E. Motte-Florac and M. Dunham. Paris: IRD, 1133–1182.

Walsh, M.T. and Goldman, H.V. 2008. Updating the inventory of Zanzibar leopard specimens. *CAT News* (Newsletter of the IUCN/SSC Cat Specialist Group), 49, 4–6.

Walsh, M.T. and Goldman, H.V. 2012. Chasing imaginary leopards: science, witchcraft and the politics of conservation in Zanzibar. *Journal of Eastern African Studies*, 6(4), 727–746.

West, H.G. 2005. *Kupilikula: Governance and the Invisible Realm in Mozambique.* Chicago and London: University of Chicago Press.

Williamson, E. 2004. *Borges: A Life.* New York: Viking.

4   The Naga tiger-man and the modern assemblage of a myth

Michael Heneise

In reality I'll be a man. I'll just take a nap [...] and then I go. In reality I'll still be a person. Only by dreaming I'll go [to the tiger world]. In the month of June and July it is the time when our souls roam around most, this is the period when I dream heavily. Not only me. All the souls become other animals in this season.[1]

Introduction

The Naga tiger-man or *tekhumiavi*[2] who uttered these words describes a phenomenon that in one form or another has revealed itself in the sculpted, painted, oral and printed histories of the world: humans and nonhuman animals take on, share or exchange attributes, as in the broad concept of theriomorphism.[3] Yet therianthropy (Greek *therion* 'wild animal' and *anthropos* 'human being', also often referred to as 'lycanthropy', with *lycos* being 'wolf' but also more generally any wild animal) has remained a subject of mythical, popular fascination; a subject hot at the Hollywood box-office, but rarely investigated in the sciences. With the re-emergence of the nature/culture debate, heightened by concerns over the effects of population growth and global warming on human and nonhuman animal habitats (e.g. Cassidy and Mills 2012; Descola 1994; 1996; 2013; van Dooren 2011; 2014; Kohn 2007; 2013; Latour 2013; Viveiros de Castro 1998; 2004), a look at more complex forms of human and nonhuman interaction is helpful in elucidating hidden truths about the inextricable life-links that bind humans, nonhuman animals, and the natural world. This chapter is an initial exploration into this phenomenon in the context of the Naga-inhabited areas in the Indo-Burmese borderlands where it is known to have existed among a few particular groups, but is virtually unknown among younger generations as a traditional practice. The reasons for its 'disappearance' are questions that require more in-depth research, though a few general working hypotheses will be presented here. These include pressures associated with population growth and the depletion of forest habitats, and the notion that puritanical forms of Christianity introduced by American Baptist missionaries have eroded traditions that accommodated therianthropy. The latter have perhaps contributed to rupturing what appears to have been a pre-Christian Naga nature/culture continuum. The emergence of a pan-Naga mythology is a more recent development, and linked to Naga nationalist assertions of difference vis-à-vis the dominant Hindu political discourse of the Indian state. Here the *tekhumiavi*, as a mythic figure with both tragic and heroic qualities, is re-emerging as a symbol that embodies the nature/culture continuum, thus representing the unity of the Naga and their vast ancestral land that underpins Naga nationalist narratives and political assertions.

For those living in the bustling urban areas of Kohima and Dimapur, the *tekhumiavi* is an urban myth, and represents the 'wildness' of the distant rural village. A young undergraduate at Baptist College in Kohima raises his arm and points somewhere to the north or the east – 'over there, the old traditions are still practised'. It is indeed in the 'north' that I first came to know about the 'tiger-man'. Chatting with a group of Ao Naga youths in Mokokchung in 2002, I noted in my notebook a series of descriptions that seemed strange at first, as they usually included the seemingly contradictory phrases 'a village nuisance' and 'a great and powerful mystery' in the same sentence. Ten years later in Kohima, most of my older informants seem uncomfortable with the subject, and often just state something along the lines 'those are old practices, but now we are Christians'. Though rumours about 'this' or 'that' person exhibiting peculiar '*tekhumiavi*-like' behaviour circulate regularly around Kohima Village (adjacent to Kohima Town), I infer that these are things one does not discuss outside of one's family or close age-group friends. However, while on a hike to the Dzülakie forest, my companion – a road contractor – shared a personal experience he attributed to the mysterious activities of a *tekhumiavi*:

> Even though we have big dogs in our compound, he reached the pens where we have rabbits and took them. I know it was a *tekhumiavi* because I have felt his presence before. You know, they have both the spirit of a man and the spirit of a tiger; and he moves like a shadow without making any noise. I felt he was following me. I don't have a problem with them, in fact I know an old Chakhesang *tekhumiavi* who lives in Kohima [...] though he was very old the last time I saw him.[4]

Despite his conviction that these were the activities of a local *tekhumiavi*, I was sceptical. This is not because I questioned the phenomenon itself, but because it seemed implausible that a *tekhumiavi* in the form of a leopard or tiger would roam into an urban area and risk being sighted, caught or worse. But the peculiar phenomenon itself felt oddly familiar.

As a youth in Nicaragua I encountered a belief in something akin to the *tekhumiavi* known as *nahaulism*. War loomed in the distance much as it does in the Naga areas, but nothing troubled young Nica imaginations more than a nocturnal visit of the *carretanahua* – a giant 'ghost-cart' driven by an evil spirit or witch, and pulled by a train of skeleton oxen. The elders' accounts often included giant black dogs that would predate on villagers foolish enough to open their doors for a glimpse (see Aragón 2006; Correa 1955; de

Guerrero 1891; Garrison Brinton 1894; Martinez Gonzales 2004; Saler 1964). As with the other neighbourhood kids in Nicaragua, I was immersed in the soup of local legends and folklore, and all this somehow elicited a combination of heightened awareness and all-consuming fear. This played well into the hands of local Christian groups keen on recruiting members. The torturous scenes of hell-fire brimstone portrayed in films projected on outdoor, makeshift screens by travelling evangelists offered a simple formula that appealed to a community burdened by war and superstition. The fear of the *nahual* – or whatever was 'out there' that sufficiently provoked a reverence for old beliefs – was invariably linked to more ultimate questions, distracting or tempering the village consciousness as casualties mounted due to the more 'temporary' political questions being played out in the mountains a few hours' drive away.

A visit to Mexico's more accessible archaeological sites reveals myriad therianthropic images – in particular that of the jaguar-warrior (see also Benson 1998; Miller and Taube 1993; Tate 1999). As mentioned earlier, traces of *nahualism* are found in contemporary mythology (Martinez Gonzales 2004; see also Fuentes and Guzman 1882), and have been studied extensively by anthropologists in Central Mexico, Chiapas and the Yucatan since the mid-1800s (see Garrison Brinton 1894; Kaplan 1956; Nutini and Roberts 1993; Stratmeyer and Stratmeyer 1977). Hutton, the first to describe the Naga *tekhumiavi* at length (especially 1920 and 1921b: 200–208), suggests Naga concepts and beliefs about the *tekhumiavi* are unique in the region, but also notes they closely resemble Meso-American *nahualism*. As with *nahualism* in the Americas, re-imaginings, embellishments and circulation of the myth of the *tekhumiavi* in urban legends and folktales sustain its popular fascination.

The Naga *tekhumiavi* elicits similar reactions to the Meso-American *nahual* from younger generations in Kohima, though very few know it was once a more widespread practice and belief, and that it may be still traced in some areas of Nagaland.[5] Preliminary research on the subject of tiger transformation may be found throughout Southeast Asia, with some of the most detailed ethnographic work on the subject focused in Malaysia and Indonesia (see Bakels 1994; Boomgaard 1994; Endicott 1979; Wessing 1986).[6] In Northeast India the phenomenon has been recorded among a variety of large indigenous groups and sub-groups including the Khasis, Garos, Chins and the Nagas. Though there is really no standard definition or way of counting the various communities belonging under the Naga designation,[7] the focus here is on a few larger groups that have historically been associated with the practice among the Naga, namely the Ao, Chakhesang, Chang, Khiamniungan, Konyak, Lotha, Sangtam, Sumi and Yimchungrü communities.

Instructive in this inquiry is that the widespread belief in therianthropy found among these groups and in the wider Southeast Asian region is accompanied by an ubiquity of creation myths giving reverence to, and reflecting both a fascination and an affinity with, powerful feline predators such as leopards and tigers (see Abbot and Thant Han 2000; Elwin 1958; 1970; Norbu 1999). An ancient Ao Naga myth reflects this:

in the beginning of the earth there was no distinction between light and darkness [and] men and animals lived together in perfect understanding and harmony [...] [thus the spread of] tales of girls being 'married' to tigers and lovely maidens having trees as lovers.

(Ao 1999: 65–66)

Similar myths of origin are found among Native Americans, in Amazonia (see for example Kohn 2007 and Viveiros de Castro 1998) and among the Inuit of Alaska, whose traditions 'describe an early period when all animals and humans lived in the same community, speaking the same language, frequently changing appearance, and intermarrying' (Willis 2006: 31). Students of mythology would perhaps find these passages unremarkable, and indeed in nearly every way they adhere to universal ideas and symbols found in creation myths throughout the world (e.g. Lévi-Strauss 1964–1971). Yet the metaphorical quality typically associated with such origin myths is unsettled by accounts documented by ethnographers that are more consistent with the literal approach found in local understandings. Illustrative of this are ethnographies detailing the meticulous regulation of various food-related taboos, hunting practices, and other rituals invested in the notion of a common kinship between humans, nonhuman animals (specifically tigers or leopards) and spirits. Hutton (1921a: 208), for instance, writes:

When an Angami village kills a tiger or a leopard the Kemovo (priest or elder) proclaims a non-working day for the death of an 'elder brother'. The flesh of tigers and leopards is often eaten by Angamis (men only and under certain restrictions), that of leopards (never of tiger) by the Changs, but the Sema would not dream of eating either. It is absolutely *genna* (taboo) to touch it, and most Sema villages, if they kill a tiger or a leopard, leave the body to rot where it lies, though the head may be taken and brought back to the village. The fear of the tiger among all Nagas is considerable, and all regard them as beings apart from the ordinary wild animals and very nearly connected with the human race. Thus a man who is descended from one who was killed by a tiger will not eat meat from a tiger's kill, as it would be equivalent to sharing the dish of an hereditary enemy.

Ao (1999; see also Changkija 2007) suggests that this belief in the common kinship between humans and nonhuman animals extends to the notion of soul co-essence, and that this is a significant idea governing Naga ontology, namely 'the human soul can reside in forms other than the human' (1999: 66). Jacobs, citing W. G. Archer, records the words of a Sangtam Naga *tekhumiavi* describing the process as a shifting of the soul between the human body and that of the animal:

My soul does not live in my body. It lives in the leopard. It is not in me now. It visits me in sleep. I meet it in dreams. Then I know what it [sic] has been doing … If anything happened to my leopard in the day, my soul would come and tell me. I would get the same wounds.

(Jacobs 1998: 85–86)

As with the testimony I included at the introduction, the Sangtam *tekhumiavi*, at the time of transformation is engaged in a kind of soul-transfer with a leopard, where the man's body remains untransformed, and the man's spirit or soul is channelled in a sort of deep dream state and becomes fully engaged in the actions of the nonhuman animal. The body of the man (who may be lying in bed at home) moves convulsively and shifts around as the tiger or leopard moves about in the forests or fields outside. Entering the tiger or leopard during sleep, the soul usually returns to the human body with daylight. But if, for example, it were to remain in the nonhuman animal for a period of days, the human body, though conscious, would be weak and lethargic. Hutton relates; 'During sleep the soul is the leopard with its full faculties, but when the human body is awake the soul is only semiconsciously, if at all, aware of its doings as a leopard, unless under the influence of some violent emotion, such as fear, experienced by the leopard' (1921b: 202). According to Hutton, *tekhumiavi* possessions are often accompanied by swelling and severe pain in the knee and elbow joints, as well as pain in the small of the back, depending on the level of activity of the tiger or leopard (1921b: 202).

Regarding the acquisition of the unique ability, most accounts suggest it happens at an early age. Detailing the Sumi Naga beliefs on the subject, Hutton's informants suggested that the phenomenon could not be induced by external means but that it could be acquired through close association with a known *tekhumiavi*. The ability could also be obtained by being fed 'chicken flesh and ginger … given in successive collections of six, five and three pieces of each together on crossed pieces of plantain leaf' by a *tekhumiavi* (1921: 201). Among the Ao Nagas, the belief is that the ability is inherited from either the paternal or maternal lines, though a generation gap may occur in the line of succession. In addition, multiple family members with a tiger-spirit are not known to exist or to have existed. Inherited or non-inherited, pre-destination is necessary and those possessing a tiger-spirit can discern if others are also predestined to receive the 'gift' (Sutter 2008: 280). Still others suggest that the tiger-spirit can also be acquired by appealing to a person who is 'reputed to be in possession of such a power'. Ao writes that 'if the applicant's prayer is to be granted, he will be offered a cup of wine or pipe to smoke by the person to whom he has appealed' (1999: 67).

In an interview I conducted in 2010, my informant – a former soldier in the Naga underground in the 1960s and 1970s who claimed to have personally known many *tekhumiavi* – was eager to explain the process in which a person acquires the tiger-spirit:

In the beginning, the [tiger-spirit] takes the shape of small insects, it then takes the form of a butterfly and the like, progressing into […] the shape of animals like squirrels, and rats. Later on it is growing, getting bigger and [...] it will change into animals like dogs, in physical appearance. When that happens, it will start attacking livestock. And it will eventually turn into a tiger. When it turns into an actual tiger, the person's tiger-spirit will roam around people's houses, properties and will start attacking chickens. The person's tiger-spirit not only attacks and eats the chickens, but it will start scaring people or show his tiger-spirit to others.

Descriptions of the world that the *tekhumiavi* navigates in tiger form are rare. Though there is a considerable gap in anthropological research in the Naga-inhabited areas from the 1940s through the 1990s due to the Indo-Naga War, recent fieldwork on the subject (most notably Ao 1999; Longchar and Vashum 1998; Longchar and Davis 1999; Longchar 2000; Sutter 2008) also demonstrates striking similarities between current and early descriptions of the tiger-spirit world (particularly Hutton 1920; 1921a; 1921b; 1931; 1942; Mills 1922; Smith 1925), namely that it is socially organised in ways that resemble human social structure. Nonhuman animals in this domain are social beings – with societies, political structures, and specifically designated duties, the distribution of which is based on a stratified system of ranks. Informants who have participated in this world in their transformed tiger forms describe the existence of tiger conferences, tiger battalions, and companies with rules and laws – all dictated and enforced by a 'council of tigers':

> On 5th August we used to have a conference, after the harvest … every year. [...] Piyong Tenem is near the Dikhu riverbank. There is one waterfall, a very big one. This is the place. [...] There are also seasons … We use to decide in full moon nights. During full moon we hold our meetings. [...] And then what kind of animals we can take – all these things are discussed. We have to divide the animals among ourselves. [...] And then, there is also a discussion about the fields of the farmers. We have a system. We will discuss about the farmers. Whether the next harvest will be good or bad – we will decide. If the farmers face water problems we will call the water. [...] But if they are in bad luck, they use to face problems from us. Means: It depends upon the mood of the Water God.
>
> (Sutter 2008: 272)

The laws set forth by this underworld encompass a wide range of activities for the tiger-men, including a prescribed diet. The spirit-guardians indicate which animals can be hunted, and as different spirits have 'ownership' of different areas of the forest, mountain and rivers, they must be consulted before a kill (Sutter 2008: 272–273). Besides the arduous tasks demanded by the 'council of tigers', and the demands of the spirits, interviewees also described a lighter side. Sutter writes:

[It is] a complete tiger world in which they, as tigers, make jokes, fight, marry and sometimes even have families. The Ao tiger-man from Khensa even sang a song in a fictive language, which male and female tigers sing out loud in call and response form: 'Every time we meet, the male tigers used to hold their *daos* and then they sing: *Ejemtachuyula, ejemtachuyula* and the females will respond to them: *Ankangkangrakshawai, ankangkangrakshawai*'.

(2008: 285)

Tekhumiavi would often spend days at a time in their darkened homes, asleep and engaged in soul-travel, living and interacting with other spirits through the bodies of animals which may be hundreds of kilometres away.

Eduardo Viveiros de Castro's 'perspectivism' thesis (1998) draws from similar findings among Amazonian indigenous peoples. In these perspectival ontologies, animals and spirits are anthropomorphic, and though masked by bodily 'skins', like humans, they are socially organised and can possess agency. Along similar lines, Bird-David (1999) in her efforts to recast the term 'animism', develops the idea of 'relational epistemologies' wherein human beings co-inhabit the world alongside other people, some of whom may not be human. This of course recalls our earlier discussion regarding human and nonhuman animal kinship in the creation myths of the different Naga groups. Notions of personhood, therefore, are developed by human and nonhuman beings through shared, embodied interactions and necessarily maintained in relation to each other.

We might view this nature/culture continuum as necessitating a transfer of knowledge between the two, and this is usually done by mediums; individuals endowed with abilities that enable them to see the continuum and provide necessary insight to those who have greater difficulty with such perception. Oral tradition, of course, is just as important in this transfer of knowledge as the practices of skilled or gifted individuals (healers, shamans, mediums, etc.), and folk-tales and myths shared among community learners deliver a tried-and-tested classification system for carefully discerning the multitude of visible and invisible signs and actants one encounters; a map of this vast hidden domain and a prescribed set of behaviours and rituals one must adhere to in order to traverse the terrain without provoking disequilibrium and ultimately misfortune on a community.

Easterine Kire (alias Iralu), a distinguished Naga poet and writer, has established a pan-Naga literary canon by exploring these concepts as they are common among the various Naga traditions (Iralu 2001). Here the attempt is to employ a familiar framework that translates beyond the Angami Naga pre-Christian traditions of her own ancestry. Moving up in this space, however, is incrementally more difficult, as the various Naga traditions would populate this general framework differently given the distinct names, characteristics and behaviours of their deities, spirits, hierarchies and domains. Though her own poetic license to massage these differences is operative (and somewhat expected), there are additional repositories from which she draws material that resonates with her readers, namely Christian theology, and to a lesser extent politics.

It must be stated at this juncture that the *tekhumiavi* phenomenon is not practised among the Angami Nagas, nor has it been recorded among them. Hutton's monograph (1921a: 243) states this quite plainly: 'Lycanthropy is believed in but not practiced by the Angamis, though their neighbours and perhaps near relatives the Semas are inveterate lycanthropists.' This is important because Kire's writing on the Naga *tekhumiavi* (or as she prefers 'tiger-man' or 'weretigerman') is situated within a recognisable Angami context and cosmology, thus constituting an amalgamation of distinct Naga traditions (Iralu 2001).[8] Before proceeding to her text, however, a general overview of the Angami belief system will be helpful.

Kepenuopfü, literally 'birth spirit', is believed to be both the creator of all living beings, as well as ancestors of humans. As the aforementioned origin myth places 'Man', Tiger (and all larger cats) and Spirit with a common ancestor (though Hutton is not explicit), *Kepenuopfü* may be placed as one-and-the-same with the deceased mother now ascended in spirit form.[9] All other animals were brought by the *terhuomia*, who are believed to be lower deities, chief among them being *Rutzeh*, who is evil and for whom such things as sudden death are blamed. There is also the mischievous *Telepfü* who is held responsible when men, women and children go missing. It is believed that *Telepfü* steals her captives of their senses so they are unable to scream or run, but she does not kill them. Benevolent *terhuomia* include *Maweno* who is known to visit farmers, and blesses people with good crops or livestock; *Ayepi*, a sort of fairy who lives with humans in their homes and brings prosperity; *Tsükho and Dzürawü*, male and female spirits represented as dwarfs, are believed to live in the forests and are guardians of the wild animals; and *Metsimo*, who, as Hutton writes, 'guards the approach to paradise, a sort of Angami St. Peter', possessing attributes (according to the Memi Tenyimia tradition) that could qualify him as *Kepenuopfü's* husband (1921a: 182 and footnote). There is also a belief in *Rhopfü*, a female spirit that is attached to or accompanies every human being. According to Hutton, *Rhopfü* is:

> a mysterious spiritual force which seems to combine the attributes of guardian angel, familiar spirit, Destiny, and in some cases it would seem even of a man's own soul. The description is vague enough, but the danger to be avoided in transcribing any Angami ideas upon the supernatural is, above all, the danger of distinguishing what is vague, of giving form to what is void, of defining what is not finite.
>
> (1921a: 183)

Tekhu-rho, the god and guardian of tigers and leopards, is considered a very powerful deity, and believed to be responsible for the disappearance of humans (primarily of hunters) in the jungle. At the moment of a disappearance, village elders will declare a *genna* or 'prohibition' (in this case entailing a community-wide freeze on field labour) in an effort to appease the angry deity. Search parties will then enter the forest to begin a search. There is a tradition among many Naga groups involving the propping open of the jaw of a killed tiger with a bamboo or wooden stick and placing its head into a running stream far from the village. The belief is that, if prepared quickly enough, when *Tekhu-rho* asks the dead tiger for the name of the responsible hunter – seeking to avenge the killing – he will only hear the meaningless gargle of water (Hutton 1921a: 182). *Terhuomia* are potentially countless, every village having a variety of *terhuomia* that a neighbouring village may not have heard of. The hundreds of myths, legends, folktales and songs that are still shared around kitchen hearths are firmly contextualised within this hidden world where *terhuomia* inhabit stones, trees, forests, rivers and mountains. According to one of my informants, 'the spirits demand respect and enact harsh punishments on those who are careless and don't follow the ancestral laws'.[10]

Angami beliefs about the *tekhumiavi* straddle two worlds – the world of human, waking reality, and a very different, seemingly mythological one – a vast universe of spirits, signs and omens (see Changkija 2007 for a similar description and beliefs among the Ao). This invisible world functions as a repository of knowledge, housing historical, religious, and moral teachings accessible through dreams or in the telling of folktales, in the singing of songs, in the enactment of rituals, and in some cases through divination (see Iralu 2001).[11] In her book *Naga Folktales Retold* (Iralu 2009: 75) Kire situates the *tekhumiavi* in relation to family and village life, and in relation to larger historical imperatives:

> He was half-man, half-spirit, the last in a long line of weretigermen. Tsaricho's father and his father before him had carried the spirit of the tiger in their beings. Many in his family thought he would bequeath this strange legacy to his young son who was unformed but not much younger than Tsaricho himself when he became a tiger-man. But he would have none of it. The lad attended the Mission School and sang the songs of the Lamb. It was not long before Tsaricho could tell that the boy did not have it in him to carry the spirit of the tiger. To his credit he refrained from imposing his will on his only son. He checked the elders with the words: 'Let it be, his is a destiny different from ours.' Was he seven or eight when his father had called him to his side of the fire and shared portions of chicken liver and fragments of country ginger with him? He could not remember now but he could certainly remember a time when he had not carried the spirit of the tiger.

In this passage, Kire is careful to describe Angami beliefs about tiger-spirit transmission in which a person either inherits or gains the tiger-spirit through the sharing of pieces of fresh ginger root and chicken liver with a known *tekhumiavi*. The reader also gets a sense of the level of *terhuomia* activity present in the forest. In the tale, two men from a neighbouring village have just threatened to shoot Tsaricho's great grandfather's tiger if they see it near their village, and are now returning home (Iralu 2001: 76):

> The men turned to go but not before glimpsing the unspoken anger on the old man's face. It was this memory of his great grandfather that returned to Tsaricho – the sight of his countenance darkening like a rapidly gathering storm sky. The two men never reached their village. They were on the thin path that intersected the woods when they heard the rumble of thunder. They turned this way and that but all around them trees crashed down uprooted and upclawed. The first man opened his mouth to shout but words failed to come: in the next moment his blood froze within him as he saw the giant figure of a tiger leap out of the darkness. Before the tiger descended upon him, his last thought was that he had never seen a more magnificent animal before, the symmetry of blood and sinew wonderfully displayed in the mid-air leap.

Kire describes a convergence of invisible agents involved in orchestrating the attack, capturing the men who would defy the *tekhumiavi* (who at this point is almost equated with a demigod, as he is both spirit and tiger, and commands the invisible powers of the forest who participate in the killing). Tsaricho's great grandfather bridges the continuum between visible and the invisible and occupies a powerful, yet sacrificial element in the story, reconciling the brokenness that prevails in the mythological roster between Man and Tiger, but ultimately suffering classic 'heroic' consequences, as he is feared and thus marginalised from the community. Undoubtedly, the text integrates more recent ideas with old beliefs and thus creates a genre that ties into the mythological with a reinterpretation that will appeal to modern readers. One significant element in the story is the liminal state of the *tekhumiavi* vis-à-vis the community. By all accounts, this is a recent development that has precipitated the disappearance of the *tekhumiavi* from the community. This marginality in relation to the community is picked up by Sutter's informant as extending to issues of practicality (2008: 271):

> My tiger used to come up to the Milak river. He very much wants to come to my place, but my *chubok*, the space below my kitchen, is too small … there is not enough space here for him to sleep […] He is seven feet long, my tiger […] That's why. Nowadays he never comes, he never crosses the Milak river. […] Only sometimes, very rarely, he even comes to my place. But it is very difficult since the villagers are afraid of tigers. I don't want to disturb the villagers. That's why I don't allow him to come here.

Kire's story (Iralu 2001) also introduces, in a subtle way, the influence of early Christian missions on the village community – particularly the introduction of an English-medium mission school for primary education. This more or less situates the story at the junction of two belief systems, and as alluded to earlier, the text reveals elements of Christian theology – particularly soteriology – and hints at some compatibility between them.[12] Nevertheless, a once revered village *tekhumiavi* is being displaced, and though he must make a choice regarding the continuation or cessation of a predestined and in some sense, a sacrificial legacy, his fate has been decided. The author reveals this at the very outset of the story, and this may very well be suggestive of lamentation.

As discussed throughout, the most significant research on the subject of Naga therianthropy was conducted nearly a century ago by Hutton (with important publications on the subject in 1920 and 1931), and to a lesser extent Mills (1922; 1926). Their overall research agenda followed a predefined outline conceived in response to a general concern that the colourful Naga traditions were vanishing and needed urgent documenting.[13] But their work in relation to the *tekhumiavi* phenomenon went over and above this requirement. Clearly engrossed by the subject, they collected anecdotes, documented personal experiences and aimed to analytically clarify the mechanisms employed in the enactment of tiger-transformation (the idea of the transferability of a soul or spirit between hosts), and investigated at length the parallels they observed between Naga tiger-transformation and *nahualism* in the Americas. It is no stretch to imagine that it was fuelled by an intense personal curiosity. What is lacking, however, is a more detailed recording of the oral histories of the various Naga groups, and how practices such as tiger-transformation fit within them. In this regard, the project was left unfinished or open-ended, and it will be up to a new wave of anthropologists and scholars concerned with recording the oral histories of the Nagas – led to some degree by Ao (1999), Longchar (2000), and more recently Sutter (2008) and Kaiser (in Oppitz et al. 2008) – to pick up those threads and take the research further. This brief summary of Naga tiger-transformations including some recent interviews with Naga informants will hopefully contribute to the cause.

Conclusion

Research into this unique and complex phenomenon is considerably more difficult now than when the scholars whose work I have summarised above were active. For one, deforestation, over-hunting and poaching have increased at alarming rates in the past few decades, and many believe the extinction of large feline predators such as tigers is only a matter of time. Research restrictions have also been in effect since the advent of the Indo-Naga war in the 1950s, and it is only since 2011 that visitors have been allowed to stay longer than the ten days prescribed by the once strictly enforced Restricted Area designation. A further concern is the extent to which Christian institutions have managed to subdue traditional religious practices that offer glimpses into vast hidden spaces that remain largely unexplored. Finally, inter-generational transmission of oral history has been severely quieted, and though it is difficult to identify any one cause, the reshuffling of community space, state-led education reforms and the general perception among older generations that young people are disinterested in their traditions all seem to contribute to the problem. However, these are areas that require further investigation.

One approach is to follow the local narrative as closely as possible. For instance, one early Saturday morning in August, I made my way down through Kohima Village to a house on the eastern slope for a cup of tea with a family I had met some months back before the monsoon. They asked what I was working on currently, and I mentioned the research on the *tekhumiavi*. My hosts were quite surprised to hear this. 'You know of this?' I explained that I was writing a general overview of what had been recorded on the subject by earlier researchers, with hopes of opening a new project on contemporary perceptions. The woman spoke first. 'There is a *tekhumiavi* near our fields. We saw his tracks a few months back. These are very large, and the rains didn't even wash them away – they have five-fingered claws.' Her husband then spoke, stating '*tekhumiavi* tracks have five-fingered imprints – not four.'[14] I asked where their fields were, and she pointed in the direction of the Border Security Force encampment on the road to Phek. Not terribly far from the BSF camp lives a good friend of mine, the road contractor cited above, whose account I had initially questioned, but which I now have to accept I may have written off too soon!

Notes

1 Quote by NizhatoZhimomi of Khukyie village, Nagaland; interviewed by Sutter (2008: 286).
2 The term *tekhumiavi* (literally 'tiger grown into the shape of') is an Angami Naga term used as a general description of the phenomenon and typically understood to be inclusive of the experiences of women as well as men with this particular ability. *Theku* or 'tiger' refers certainly to the Bengal tiger (*Panthera tigris tigris*) but is also the term used for the much more common leopard (*Panthera pardus delacouri*), and smaller cats such as the golden cat (*Felis aurata*) in the Naga region. That said, *tekhumiavi* would also be used to describe rare occurrences involving other animal hosts such as monkeys, dogs, and snakes.
3 The Hindu concept of *avatar* is worth mentioning here as it shares some characteristics with the idea of soul-transfer or what is often referred to as shapeshifting. In this belief a deity descends (in Sanskrit *ava* is 'descend', and *tar* is 'to cross') and manifests itself (i.e. 'incarnates') in human bodily form. Though there is little relationship between Hinduism and the belief system of the Nagas, it is nevertheless indicative of the ubiquity of this idea of soul transference or astral projection.
4 Interview with A. Chielie in Dzülakie, January 2010, Nagaland, India.
5 These accounts are drawn from a series of interviews conducted among mostly teenagers in Kohima in January and February 2013.
6 In November of 2009 I interviewed Dr Gam Shae, a Baptist missionary who was serving in Banjarmasin, the provincial capital of Kalimantan, Indonesia in 1998 when he witnessed a very peculiar event in his own home. A guest student, Henry, 'was on the floor again growling, gnashing his teeth, clawing the floor, resisting anyone who restrained him with superhuman strength. Henry did not know what transpired during this type of experience. But he knew that all of his friends

became afraid of him and rejected him for friendship. As a member of his own cultural group, he shared with his friends the understanding that he was under the influence of a power outside of himself. He believed that he was possessed by the spirit of his ancestor called the tiger spirit. Furthermore he was told by his relatives that whenever he was possessed by that spirit, he would become lucky. So while it was a frightening experience for his friends, he looked upon it as a positive experience that would bring fortune.'

7 Some suggest that there exist anywhere from thirty to just under eighty tribes belonging to the larger Naga group (Oppitz et al. 2008: 16).

8 The *tekhumiavi* phenomenon itself is an adoption from Angami neighbours – likely the Chakhesang, close neighbours of the Angamis. The Chakhesang Nagas are in fact a recent clustering of three Naga groups – Chakri, Kezha and Sangtam – all members of the larger Tenyimia cultural-linguistic group which includes the Angamis. However, the distinction is made as the Angamis likely split off from the former some four or five centuries ago (taking the age of Angami villages as a measure).

9 This follows my own deductive reasoning based on Hutton's statement: 'The dwelling place of Kepenopü is always located in the sky, and the souls of those who have lived good lives, according to the Angami standard, that is, go to the sky after death and dwell with her' (1921a: 181). The Sumi Nagas also believe in a Supreme God named *Alhou or Timilhou*. The Sumi also believe in spirits of the sky, *Kungumi*, which can be equated with Tenyimia concepts of *terhuomia* (see Hutton 1921b: 191).

10 My informant, the former soldier of the underground, speaks of river spirits, forest spirits, and mountain spirits with complete candour and seriousness. He spoke at length about the daughter of *Japfü* (the third-highest mountain peak in Nagaland). *Japfü*, angry with his daughter, cursed her by splitting her head down the middle (a mountain near Wokha has a large slit down the middle from which a beautiful waterfall spills out). The wife of a man from Dimapur was reputed to have had a strange relationship with the river spirit which was said to be *Japfü's* daughter. She would spend half of the year with her human husband in Dimapur, and the other half with her female spirit-lover in Wokha (15 January 2010).

11 Unfortunately, much of this material has yet to be recorded. Stuart Blackburn, in *Naga Identities*, writes: 'Naga oral traditions have not been adequately researched. We have a reasonable body of folktales available in print, but little in the way of other texts, and even this corpus of tales largely rests on field work done nearly a century ago. Since those days, little work has been carried out on oral traditions amount the Nagas' (Blackburn 2008: 260).

12 Discussions I have had with numerous Naga theologians and many other informants over the years confirm a general belief that Naga traditional religion was not dissimilar to the Hebrew traditions of the Old Testament, and thus 'pre-fitted' for the introduction of Christianity.

13 In the Naga areas, Hutton's 1921 monograph *The Angami Nagas* was the first volume to be published following what was by then a standardised ethnographic outline developed by the Assam Administration, using as its model the 1907 monograph on the Khasis of Assam written by Major P. R. T. Gurden (Deputy Commissioner of Eastern Bengal and Assam, and Superintendent of Ethnography in Assam). Gurden had been charged with supervising fifteen such monographs that were to cover the larger indigenous groups found in the Patkai mountains (Naga Hills); the plains and surrounding hill areas of Manipur (Lushai and Kuki groups); and the southeastern areas of Assam (Lushais) including present-day Meghalaya (Garos and Khasis). J. H. Hutton would revive the series after stalled activity due to the First World War (Vidyarthi 1979: 54–60).

14 Interview with A. R. family, P-Khel, Kohima Village, 10 August 2013.

References

Abbot, G. and Thant Han, K. 2000. *The Folk Tales of Burma: An Introduction.* (Handbuch der Orientalistik III, vol. 11). Leiden/Boston/Cologne: Brill.

Ao, T. 1999. *Ao-Naga Oral Tradition.* Baroda, India: Bhasha Publications.

Aragón, E.A. 2006. Tradición oral y representaciónliteraria: El mito nicaragüense de la Carreta Nagua. *Literatura, crítica y comentarios* (Accessed 18 August 2013 from http://erickaguirre.blogspot.in/2006/07/tradicin-oral-y-representacin.html).

Bakels, J. 1994. But his stripes remain: on the symbolism of tiger in the oral tradition of Kerinci, Sumatra. In *Text and Tales: Studies in Oral Tradition*, edited by J. Bakels and J. Oosten. Leiden: Netherlands Research School CNWS, 33–51.

Benson, E.P. 1998. The lord, the ruler: jaguar symbolism in the Americas. In *Icons of Power: Feline Symbolism in the Americas*, edited by N.J. Saunders. London: Routledge, 53–76.

Bird-David, N. 1999. 'Animism' revisited: personhood, environment, and relational epistemology. *Current Anthropology*, 40(SI), S67–S91.

Blackburn, S. 2008. The stories stones tell: Naga oral stories and culture. In *Naga Identities: Changing Local Cultures in the Northeast of India*, edited by M. Oppitz, T. Kaiser, A. von Stockhausen and M. Wettstein. Ghent: Snoeck Publishers, 259–270.

Boomgaard, P. 1994. Death to the tiger! The development of tiger and leopard rituals in Java, 1605–1906. *South East Asia Research*, 2(2), 141–175.

Cassidy, A. and Mills, B. 2012. 'Fox tots attack shock': Urban foxes, mass media and boundary-breaching. *Environmental Communication: A Journal of Nature and Culture*, 6(4), 494–511.

Changkija, N. 2007. From oral tale to graphic novel: re-animating the tiger-soul. Ph.D. dissertation, Faculty of Arts, Griffith University (Accessed on 18 August 2013 from http://www4.gu.edu.au:8080/adt-root/public/adt-QGU2009218.072040/index.html).

Correa, G. 1955. El espíritudel mal en Guatemala. In *Nativism and Syncretism.* Middle American Research Institute Publications 19. New Orleans, LA: Tulane University Press, 37–104.

Descola, P. 1994. *In the Society of Nature: A Native Ecology in Amazonia* (N. Scott, trans.). Cambridge Studies in Social and Cultural Anthropology 93. Cambridge: Cambridge University Press (Original work published 1986).

Descola, P. 1996. *The Spears of Twilight: Life and Death in the Amazon Jungle* (J. Lloyd, trans.). New York:New Press (Original work published 1993).

Descola, P. 2013. Beyond Nature and Culture (J. Lloyd, trans.). Chicago: University of Chicago Press (Original work published 2005).

Elwin, V. 1958. *Myths of the North-east Frontier of India, Volume 1.* Assam, India: North-East Frontier Agency.

Elwin, V. 1970. *A New Book of Tribal Fiction.* Assam, India: North-East Frontier Agency.

Endicott, K. 1979. *Batek Negrito Religion: The World-View and Rituals of a Hunting and Gathering People of Peninsular Malaysia.* Oxford: Clarendon Press.

Fuentes, F.A. and Zaragoza Guzmán, J. 1882. *Historia de Guatemala ó Recordación Florida.* Madrid: L. Navarro.

Garrison Brinton, D. 1894. *Nagualism: A Study in Native American Folk-lore and History.* N.p.: MacCalla Printers.

de Guerrero, E.A.P. 1891. Games and popular superstitions of Nicaragua. *The Journal of American Folklore*, 4(12), 35–38.

Hutton, J.H. 1920. Leopard-men in the Naga Hills. *Journal of the Royal Anthropological Institute*, 50, 41–51.

Hutton, J.H. 1921a. *The Angami Nagas*. London: Macmillan.

Hutton, J.H. 1921b. *The Sema Nagas*. London: Macmillan.

Hutton, J.H. 1931. Lycanthropy. *Man in India*, 11, 208–216.

Hutton, J.H. 1942. Lycanthropy. Correspondence. *Folklore*, 53(1), 79–80.

Iralu, E. 2009. *Naga Folktales Retold*. Kohima, India: Barkweaver Publications.

Iralu, E. 2001. *The Windhover Collection*. Poona, India: Steven Herlekar.

Jacobs, J. with Macfarlane, A., Harrison, S. and Herle, A. 1998. *The Nagas: Hill Peoples of Northeast India: Society, Culture, and the Colonial Encounter*. Bangkok: River Books.

Kaplan, L. 1956. Tonal and Nagual in coastal Oaxaca. *Journal of American Folklore*, 69, 363–368.

Kohn, E. 2013. *How Forests Think: Towards and Anthropology Beyond the Human*. Berkeley: University of California Press.

Kohn, E. 2007. How dogs dream: Amazonian natures and the politics of trans-species engagement. *American Ethnologist*, 34(1), 3–24.

Latour, B. 2013. *An Inquiry into Modes of Existence: An Anthropology of the Moderns*. Cambridge, MA: Harvard University Press.

Lévi-Strauss, C. 1964–1971. *Mythologiques*. 4 vols. Paris: Plon.

Longchar, W. and Vashum, Y. 1998. *The Tribal Worldview and Ecology*. Jorhat, India: Tribal Study Centre.

Longchar, W. and Davis, L.E. (eds) 1999. *Doing Theology with Tribal Resources*. Jorhat, India: Tribal Study Centre.

Longchar, W. 2000. *The Tribal Religious Traditions in North East India: An Introduction*. Jorhat, India: Eastern Theological College.

Martinez Gonzales, R. 2004. *El Nahual y Otras Coesencias entre los Mayas: Una Primera Síntesis*. Mexico City: Instituto de Investigaciones Históricas, UNAM.

Miller, M. and Taube, K. 1993. *The Gods and Symbols of Ancient Mexico and the Maya: An Illustrated Dictionary of Mesoamerican Religion*. London: Thames & Hudson.

Mills, J.P. 1922. *The LhotaNagas*. London: Macmillan.

Mills, J.P. 1926. *The AoNagas*. London: Macmillan.

Norbu, K. 1999. A ritual winter exorcism in GnyanThog village, Qinghai. *Asian Folklore Studies*, 58, 189–203.

Nutini, H.G. and Roberts, J. M. 1993. *Bloodsucking Witchcraft: An Epistemological Study of Anthropomorphic Supernaturalism in Rural Tlaxcala*. Tucson: Arizona University Press.

Oppitz, M., Kaiser, T., von Stockhausen, A. and Wettstein, M. 2008. *Naga Identities: Changing Local Cultures in the Northeast of India*. Ghent: Snoeck Publishers.

Saler, B. 1964. Nagual, witch, and sorcerer in a Quiche village. *Ethnology*, 3(3), 305–328.

Smith, W.C. 1925. *The Ao Naga Tribe of Assam: A Study in Ethnology and Sociology*. London: Macmillan.

Stratmeyer, D., and Stratmeyer, J. 1977. The Jacaltec Nawal and the Soul Bearer in Concepcion Huista. In *Cognitive Studies of Southern Mesoamerica*, edited by H.L. Neuenschander and D.E. Arnold. Museum of Anthropology Publication 3. Dallas, TX: Summer Institute of Linguistics.

Sutter, R. 2008. Shadows and tigers: concepts of soul and tiger-men. In *Naga Identities: Changing Local Cultures in the Northeast of India*, edited by M. Oppitz, T. Kaiser, A. von Stockhausen and M. Wettstein. Ghent: Snoeck Publishers, 275–292.

Tate, C.E. 1999. Patrons of shamanic power: La Venta's supernatural entities in light of mixed beliefs. *Ancient Mesoamerica*, 10, 169–188.

van Dooren, T. 2014. *Flight Ways: Life and Loss at the Edge of Extinction*. New York: Columbia University Press.

van Dooren, T. 2011. *Vulture*. London: Reaktion Books.

Vidyarthi, L.P. 1979. *Rise of Anthropology in India: A Social Science Orientation*. Vol. 1. Delhi: Concept Publishing Company.

Viveiros de Castro, E. 1998. Cosmological deixis and Amerindian perspectivism. *Journal of the Royal Anthropological Institute* (n.s.) 4(3), 469–488.

Viveiros de Castro, E. 2004a. Perspectival anthropology and the method of controlled equivocation. *Tipití*, 2(1), 3–22.

Viveiros de Castro, E. 2004b. Exchanging perspectives: the transformation of objects into subjects in Amerindian ontologies. *Common Knowledge*, 10(3), 463–484.

Wessing, R. 1986. *The Soul of Ambiguity: The Tiger in Southeast Asia*. DeKalb: Northern Illinois University, Center for Southeast Asian Studies.

Willis, R. (ed.) 2006. *World Mythology: The Illustrated Guide*. Oxford: Oxford University Press.

5 Human predation and animal sociality

The transformational agency of 'wolf people' in Mongolia

Mette M. High

Introduction

In the mountainous region of Uyanga, there is concern about the recent surge in a kind of people who are regarded as different from other humans living in the area. Although these people are not new to the region, they have apparently never been as numerous as they are now. Whereas at first sight they look like everyone else, closer inspection might reveal subtle differences: an unusually firm stare, a sudden move of the arm, or a quick pointing of the ears. These differences are not lasting physiological features that are available for later examination and discussion. They exist only in the present moment and may disappear from view as soon as they are noticed. Locals pay much attention to the appearance of these indicators and discuss their discoveries in hushed voices. Their main concern is how to live peacefully together with a people who are considered predatory and essentially related to the wolf (*Canis lupus lupus*). More than being *like* wolves, 'wolf people' (*chono hün*) are said to transcend crucial distinctions, blurring divides between human and nonhuman realms. Whereas continuities between the human and the nonhuman have been richly documented elsewhere (e.g. Vilaça 2002; Vitebsky 2006; Viveiros de Castro 1998; Willerslev 2007), large parts of the Mongolian region place much emphasis on their distinction and separation (Pedersen 2001). Entering the nonhuman realm is considered eminently dangerous for humans and even shamans often refrain from such voyages. However, as growing numbers of wolf people are said to have emerged in Uyanga, this distinction is now becoming increasingly difficult to maintain.

In Western folkloric accounts and legends, the metamorphosis of humans and wolves usually entails the transformation of a human into a wolf (Baring-Gould 2009: 7). By shape-shifting into a theriomorphic wolf-like creature, often at times of full moon, werewolves (*lycanthrope*) attain extreme powers and can inflict harm on humans by biting their victims and potentially passing on the curse of the werewolf. On returning to their human form, werewolves are said to become physically weak and suffer intense pain (Baring-Gould 2009). In Uyanga, however, wolf people do not engage in periodic shape-shifting and their bodies remain human. Nonetheless, despite their familiar human form, they are regarded as wolf-beings. Rather than drawing on zoologists' elaborate classification of the *Canis lupus* species or the fantastic accounts of cryptozoological entities akin to werewolves, this chapter takes its analytical cue from the Mongolian association between wolves (*chono*) and wolf people (*chono hün*). By examining ethnographic and historical material on the position of wolves in Mongolian cosmology, I will argue that relations between humans and animals reveal both social continuities and moral ambiguities. In a region that has become the epicentre of a large gold rush, the search for mineral resources and the transformational agency of wolf people demonstrate the importance of moving away from a human-centred perspective on morality and personhood.

Fascination with wolves

In many parts of the world, wolves are surrounded by elaborate cultural understandings (Lindquist 2000; Knight 2003). As the predator *par excellence*, their fierce strength and intricate social behaviours have given rise to an enduring interest among human audiences. In myths and legends, wolves often occupy a prominent agentive position. At times they are responsible for creating human beginnings on earth, whilst at others they set in motion a human demise. In contemporary eco-politics, environmental advocacy groups express their fascination with the species whereas farmers and pastoralists often voice their frustration and opposition to wolves (see Lindquist 2000; Moore 1994). Irrespective of the kinds of relations that wolves have with humans, they appear to be beings who command human attention. And this is no less the case than in the Mongolian cultural region, where unparalleled popular attention has, in recent years, centred on the wolf. Selling more than 20 million official copies in the first three years of its publication, the novel *Wolf Totem*, by Chinese author Jiang Rong, has become an instant 'super-seller' (Mishra 2008).[1] Describing a man's fascination with, and intricate understanding of wolves, the novel depicts the so-called 'totemic relationship' between wolves and nomadic herders of Inner Mongolia during the Cultural Revolution. The novel has not only found a large readership and profitable market, but also drawn attention to wolves as a significant species in the region.

When Mongolians talk about wolves, they often comment on the strength, intelligence and determination of the animal. Many regard these qualities as admirable and highly desirable for both wolves and humans. Positioned as 'teachers' (*bagsh*) for hunters and herders, wolves are respected for their superior abilities in making a life on the Mongolian steppe. Although such praise and admiration for wolves may have inspired Jiang Rong's view that Mongolians 'worship' and 'follow' their 'wolf totem', my informants do not view wolves in such singular moral terms. The reverence and explicit prescription of admirable wolf qualities does not amount to a species-rooted yardstick for judging human actions. Rather than being positioned as axiomatically 'good' (*sain*), wolves occupy a broad terrain of moral evaluation.

This diversity in character surrounds wolves whether they take on their familiar body (*biye*) of a greyish coat, yellow eyes and bushy tail, or take on a different kind of physical form. Given their multiple bodies and moral evaluations, understanding the centrality of wolves in local descriptions of wolf people requires a broad investigation into their various manifestations and transformational agency.

The revered wolf

A commonly evoked reason for regarding wolves as beings worthy of unmatched respect is their association with the ethnogenesis of the Mongol people. Whereas warring groups inhabited the region for centuries, their unification and the emergence of a Mongol nation is a feat that is today attributed to Chinggis Khan (see Kaplonski 2004). As the founder of the Mongol Empire, Chinggis Khan is seen to have brought together isolated groups and transformed local enmity into solidarity and loyalty to a greater Mongol vision. According to the *Secret History of the Mongols*, Chinggis Khan was descended from the union twenty-three generations earlier of *Börte Chino* ('wolf' or 'blue-grey wolf') and *Qo'ai-maral* ('beautiful doe') (Cleaves 1982: 1; Onon 1990: 1).[2] In some chronicles these names appear as personal names for specific individuals, whereas in others they are positioned as explicitly mythological beings.[3] Scholars still debate the exact interpretive significance of these names (Onon 1990: n. 4). But rather than partaking in this debate, Mongolians commonly remark that they, as a nation, are 'descended from the wolf' and that the animal should be 'honoured accordingly'. Whether or not the old epic referred to a human or a nonhuman, a mythological creature or a historical figure, the wolf as an agentive being is attributed a foundational role in unifying the Mongols.

In Mongolia, the wolf also appears as the progenitor and guardian of important individuals such as epic heroes, influential Buddhist lamas and notable political leaders. Stories abound of how packs of wolves or single she-wolves nurtured abandoned children (Jila 2006; Sinor 1982: 238–239). Rescued and reared by wolves, the children become exceptional individuals whose names have left marks on history.[4] One such story is found in *Jangar* – the heroic epic of the Oirat Mongols. In this epic the male hero is abandoned on the steppe from the age of two and is saved by a nurturing she-wolf. She lets him suckle her milk and eventually he grows strong and courageous, ready to embark on his eventful journeys. Appearing as mythological nurturers and Chinggis Khan's ancestor, wolves are represented as powerful beings endowed with immense creational abilities.

Such creational agency is also displayed every night in the Mongolian sky. Travelling across the sky as the brightest star, Sirius, the heavenly wolf is said to 'lead', if not 'carry on its back', the sun, the moon and the wandering stars along their separate paths.[5] In his comprehensive study of Mongolian astrology and divination, Brian Baumann (2008: 270) shows how Sirius contributes to the creation of balance and order in the universe by endowing constellations with location and direction. Since all events on earth are tied to some phenomenon in the sky, the nocturnal journeys of Sirius are central to predicting the future, turning time and space into forces that are conducive to, and supportive of, human life. This important cosmological role of the wolf is reflected in expressions that refer to the wolf as a 'communicator' or 'link' between the sky and the earth. In Uyanga, for example, colloquial expressions highlight the wolf's proximity to, and collaboration with, the spirit called *lus*.[6] When referring to wolves, people call them 'messengers of the *lus*' (*lusyn zarlaga*) or 'the *lus*'s horse' (*lusyn unaa*). These expressions are not only used in esoteric and abstract discussions, but also on an everyday basis when referring to wolves. The wolf's role as a messenger for spirits is further evidenced in the widespread practice of using honorific terms such as *tengeriin nohoi* (heavenly dog) and *hangain nohoi* (dog of the mountains) rather than the common noun for wolf (*chono*).[7] If someone uses the word *chono*, it is claimed that wolves may hear the direct, offensive calling and in response inflict harm on the particular person. Since the wolf has created and every night recreates the possibilities for human life, he or she should be met with the respectful address that humans accord other powerful beings, whether human or non-human. As a progenitor of the Mongols and their universe, wolves have thus contributed to the foundations of human life positioned delicately between greater forces.

The predator

These nationalistic and respectful conceptualisations of wolves coexist alongside everyday concerns among nomadic herders about wolves attacking their herds of yaks, sheep, goats and horses. Herders talk constantly about wolves, and stories concerning packs of wolves wounding, if not killing, animals abound in the countryside. These stories are often accompanied by disgruntled remarks about the unchecked rise in wolf populations and the current government's disregard for local concerns. An elderly herder commented that 'democracy has helped the wolves, not us!' During the socialist era the government played a key role in wildlife management. It mandated countrywide wolf extermination campaigns, including 'wolf cub campaigns' held in the month of May. Wolves were officially hunted in order to keep their numbers under control and to provide furs for trade and gift-giving.[8] However, today these state-concerted efforts are no longer in place and it is now up to individuals to curb the wolf population. People are legally free to hunt wolves for private and industrial purposes without any quota restrictions or seasonal limitations, and wolf hunting has become a favourite pastime for many newly rich Mongolians. Another herder commented that 'city people come in Japanese jeeps, sometimes even in helicopters, and race across the steppe. They shoot again and again, but wolves are too clever to be caught like that.' Hunting wolves is not like hunting any other prey, and according to a Mongolian saying 'No one can kill a wolf unless it chooses to submit.' In the capital city of Ulaanbaatar, Mongolian and international wolf hunters have described this saying to me as a tribute to the skill and bravery of wolf hunters. Informants in the countryside, however, have emphasised the powerful and wilful position of wolves where their consent is needed in order for humans to be able to kill them (see also Nadasdy 2008).

Since wolves are said to bite to death as many animals as possible, herders in Uyanga take considerable precautions to prevent wolf attacks. A male household member usually stays up at night to watch over the animals. Often equipped with an old Russian rifle, the aim is to kill a wolf if sighted. Herders also take part in an elaborate scheme encouraging the elimination of wolves in the area. When a hunter kills a wolf, each family in the *bag* (smallest administrative regional unit) is expected to pay him 1,000 tögrög (83 cents[9]), amounting to about 200,000 MNT (167 USD) in total. Such amounts far exceed ordinary earnings on the steppe, and local hunters often talk enthusiastically about the prospects of killing a wolf.[10]

Wolves are not only feared for their attacks on herds, but also for their association with the *lus*, mentioned earlier. The *lus* is seen to reside primarily in the rivers and the mountains, and is usually invisible and ungendered. It can be benevolent to humans by ensuring an abundance of clean water and ripe berries, but can also become angry, especially if humans transgress taboos that inform human interactions with the landscape (*baigal'*), which, apart from the *lus*, also hosts many other spirit beings. As one informant explained:

> Trees have life (*mod amtai*). You should therefore always collect only the dead wood on the ground. If you break off a fresh branch, even by accident, you hurt the tree. It doesn't like that, so it will get upset at you. Maybe it is not that particular tree (*yag ter mod*) that will get upset at you, but the *lus* protecting the forest. If the *lus* gets upset (*uurlaval*), it's very bad for you and your family [...]

If humans disregard a taboo and upset the *lus*, diseases, accidents and wolf attacks may befall anyone living in the vicinity of where the transgressive act was carried out. As the Mongolian gold rush has unfolded, involving the frequent disregard of fundamental taboos related to the land (High 2013b), people have begun to live in perpetual fear that the *lus* and its wolf messenger might punish them.

Around the year 2000 a gold rush broke out in Uyanga, attracting more than 8,000 informal sector gold miners; that is, about four times the local population of herders. The gold rush brought growing numbers of miners to the area and problems soon emerged. Disputes arose and meetings were held in the nearby village in attempts to resolve the emerging conflicts. Locals were not only concerned about miners encroaching on common land and monopolising its resources, but also feared the consequences of miners transgressing taboos.

As miners dug into the ground in their search for gold, spirits residing in the landscape were seen to become increasingly angry (*uurlah*) and capable of inflicting calamities on people living among or near the mines. For local herders, the most common way for the *lus* to punish them was through wolf attacks. As 'the messenger of the *lus*', the speed, slyness and aggression with which wolves attacked their prey were seen as indicative of their position. The following excerpt from my fieldnotes illustrates the perceived link between wolves and spirits:

> One day my younger host brother Davaa was herding sheep and he happened to pick some wild garlic (*zerleg songino*). According to my host sister, less than ten minutes after picking the wild garlic a pack of wolves attacked the herd and killed a sheep. Just before picking the wild garlic, Davaa couldn't see any wolves at all. 'It was as if the wolves came out of the ground right when he picked the garlic. "Normal" wolves can't do that, so it was definitely because the *lus* was angry. If the *lus* is upset, wolves will appear', she said.

The image of the *lus* sending wolves from underground hinges on particular ideas about wolves that transcend their mere predatory nature. When herders talk about wolves, they invariably describe the many wicked characteristics of the animal in a slow and dramatic voice. Whereas the word 'bad' (*muu*) seems the preferred description, it is also labelled a 'thief' (*hulgaich*) who steals from others (see also Lindquist 2000: 179). Since wolves rarely devour all of their prey, people often lament that they steal without even 'needing' the stolen animal. Herders condemn such greedy theft as 'purposeless' (*utgagüi*) and entirely 'selfish' (*aminch*) of wolves. By only taking from humans and never giving anything in return, wolves are criticised for intruding into human life and destroying the wealth that humans have built up.[11] As such, wolves can be seen as *anti*-human, epitomising dangerous autonomy and careless individuality (see also Moore 1994).

Ideas about 'pollution' further consolidate the position of wolves as antithetical to human life and prosperity. By consuming the stolen goods (that is, their prey), wolves absorb all the pollution involved in such theft, which is in turn passed on to humans if they come into contact with wolves. It consequently becomes paramount for humans to avoid all exposure to the predator. The pollution surrounding wolves is referred to as *buzar*, which in daily language is used in similar ways to the word for 'rubbish' (*hog*), denoting something as filthy and disgusting. Although pollution can affect all members of a household group (*ail*), women and children are considered particularly vulnerable. Since the future longevity of the household is in this way at risk, protective bracelets and necklaces are worn, often in conjunction with birth year or Buddhist deity necklaces. As the wolf here embodies the very characteristics that undermine the reproduction of the household, it is morally evaluated as an inversion of human sociality. Feared for both their predatory hunger for herders' animals and their contempt for human life, wolves are what humans ought not to be. Contact with dead wolves, however, is an entirely different matter. Wolf pelts, adorning the hoods of urban wolf hunters' jeeps and displayed in front of rural wolf hunters' *gers* (round felt tents) after successful hunts, are not regarded as polluted but rather as aesthetically beautiful and fortune-giving. Certain parts of dead wolves are used for medicinal purposes, such as dried wolf tongue for respiratory illnesses, while other parts such as wolf paws are seen to pass on the benevolence of spirits. At the moment of death wolves thus become unambiguous for humans, finally returning the wealth and fortune that they acquired during their lifetime.

Wolf people

Although the predatory position of wolves provides a compelling moral framework for evaluating desirable and admirable human sociality, it has not given rise to stark oppositions between the human and nonhuman realms in the Mongolian gold mines. If the predatory wolf exemplifies the undesirable and manifest failures in sociality, humans are not juxtaposed as its corresponding moral opposite. Whereas ritual practices reassert distinctions and separations, the wolf people of Uyanga highlight the increasing fragility of the ontological boundaries between wolves and humans.

For the first many months of fieldwork in Uyanga, I never heard any mention of wolf people. It was a topic that was only broached after I witnessed a wolf person entering my host family. It all began with a family dispute, which took place in the early autumn when men get excessively drunk on the season's last strong fermented mare's milk and home-brewed vodka. Together with my host mother and three sisters, I had left for a one month extended migration with the weakest yaks in order to fatten them up quickly before the onset of winter. Since we were only women on the migration, we had no means to prevent a fight should a drunken visitor became aggressive. This difficult dynamic may partly explain the cautionary behaviour of my host mother in this situation, as recorded in my fieldnotes:

> We heard a motorbike approach our *ger* and it turned out to be my host mother's younger brother and one of his friends. They were both drunk and demanded to be served more alcohol. My host mother served them but tried to limit conversation. But an intense argument began between her and her brother. Eventually she walked over to the door, opened it and told him to leave. He fixed his eyes on her with great intensity but said nothing. Instead of relieving us of his presence, he grabbed hold of one of my host sisters and shouted into her face: 'What kind of family is this? What kind of woman asks a visitor to leave? Are you all idiots?' My host sister started crying and my host mother grabbed hold of him and shouted furiously at him: 'That's enough! Leave our place now! Now! What has happened to you? Look at yourself! You are always drunk, you do no

work, you show no respect. Leave!' He let go of my host sister, jumped over to our burning stove and kicked it with full force. The teapot placed on top of the stove landed on the ground, spilling tea all over the floor. The flimsy metal chimney attached to the stove broke off and the *ger* was soon filled with thick smoke. My host sisters covered their eyes and ears whilst crying out bits of Buddhist mantras. The youngest ran over to the altar, lit butter candles and incense, and spun the prayer wheel repeatedly. But he was not done yet. He grabbed my host sister again and shouted into her face: 'Wolf (*chono*)! You are a wolf! I know you are!' He then let go of her, ran out of the *ger* and left with his friend as my host mother shouted: 'I will never see you again! Don't ever come back to us!'

By destroying the stove, the man angered the Master of Fire (*galny ezen*) who resides in the stove and attacked a concrete manifestation of the lasting nourishment and happiness of the 'patriclan' (*ovog*).[12] In this way threatening the stability of the group, he made a powerful statement that was encapsulated and intensified in the dreaded and very rarely used swear word *chono*, conveying that my host family was behaving with wild autonomy and dangerous greed.[13] But in contrast to other swear words, the labelling of someone as *chono* goes far beyond the more usual lexicon of debasing comments. It is an identification that not only comments on how you are seen to behave, but also what kind of being you are perceived to be. It is an identification of essential difference, marking the person as having become categorically distinct from others.

Several years after the dispute, my host sister was still worried about having been called a wolf by her maternal uncle. She asked for repeated reassurance from her kinsmen that, although wolf people had an increasing presence in the area, she was not one of them. Despite living near the mines, her household group rarely took part in the mining and this was, for her, evidence that she had been wrongly denounced. She decried the kind of life that she would be destined to have if she really was a wolf person. Rather than marrying, having children and becoming part of her affinal household group, she described to me how she would be destined to live alone for the rest of her life, only inter-acting with others through her greed and thieving. In this way endangering those humans she came into contact with, she would be a predator and all others would become her prey. Apart from threatening the peaceful conviviality of others, she would prevent the balancing of forces in the landscape (*baigal*). By not respecting other human and nonhuman beings, she would ultimately put everyone's potential for peaceful and productive living at risk.

Destined for such a solitary life, wolf people cannot have offspring and as a result they only pose a risk during their own lifetime. They can never have descendants and pass on their wolf-being to future generations. In turn, they are not regarded as inheriting their wolf-being from previous generations, not even in an indirect line. Although ideas about reincarnation strongly inform local conceptualisations of human relatedness, wolf people are thus not viewed in such terms. Rather, these human predators are described as emphatically generational and corporeally contained, revealing their wolf-being only through personal actions in the present.

Whereas people whose bodily exterior conceals a wolf-being are said to have a deep history in Uyanga, today there appears to be a growing abundance of them. As thousands of people enter the mines and pan for gold in the streams, the *lus* is becoming increasingly upset with the actions of humans and punishes humanity by sending its messenger, the wolf. At a time when herders remark on the burgeoning wolf population and criticise the government for not trying to keep the predators in check, distinctions between humans and nonhumans are becoming increasingly blurred. In people's search for gold, they assert a predatory relationship with both other humans and the *lus*. Finding themselves in a cosmological struggle with spirits and wolves, questions of intentionality and morality appear ever more crucial for the preservation of everyday conviviality as well as local understandings of human and nonhuman relations in Uyanga.

Conclusion

Following the advent of the gold rush on the Mongolian steppe, wolves have become increasingly common in both human and nonhuman domains. At times they make their appearance as feared predators sent by the *lus* to prey on livestock. At others, they mingle with humans, physically similar to other humans, but destined, by virtue of their wolf-being, to become immoral recluses. Regardless of their particular manifestation, wolves possess powerful capabilities that affect humans in the most fundamental ways. Capable of stealing animals, polluting people and taking on human bodily forms, wolves can strongly influence the material prosperity and survival of local households. However, wolves are beings that do not lend themselves cogently to analytical attempts at singularisation and categorisation. It would be a gross oversimplification to cast wolves in terms similar to those used by herders when they complain about wolf attacks. Wolves are not simply 'bad' (*muu*), stealing animals at a pace that no human thief can match. Admired for their strength, intelligence and determination, wolves are also highly esteemed beings who are often honoured through depictions, gift-giving and linguistic practices. As central figures in both mythological accounts and the popular historical consciousness of Mongol origins, wolves are more than feared predators. They possess a unique agency among animals. For example, when a person is in serious need, a lone she-wolf may come to the rescue; when in the wrong, an angry wolf may inflict his punishment. Whether or not the wolf punishes or rescues, destroys or creates, he or she is capable of affecting human lives in fundamental ways.[14] Just as Sirius travels across the sky every night in his recreation of cosmological order, wolves also travel across the steppe in pursuit of potential prey. The wolf is a powerful force within the landscape that demands human attention. More than a mere currency for representation, the wolf is a being of multiple agentive forms.

In the Mongolian context, the ability to transcend boundaries and destabilise discontinuities is a rare ability associated specifically with shamanic processes (Humphrey 1995; Humphrey and Onon 1996). According to Morten Pedersen (2001), the Mongolian region is dominated by a form of 'totemist differentiation'; one that emphasises comparative differences between humans and nonhumans. In this 'heterogeneous conglomerate of mutually independent domains inhabited by humans as well as nonhumans' (2001: 418), the world appears as a 'grid' of strict boundaries and discontinuities. This grid, Pedersen argues, is contingent upon the more general vertical organisation of society evidenced by a hierarchical ethos, inherited leadership and patrilineal descent. The ability to circumvent hierarchies and travel across ontological boundaries is not an ability available to most beings, but rather a highly specialised and restricted ability surrounded by much fear and danger. Whenever shamans, the experts in metamorphosis, undertake such journeys, the preferred perspective is thus still human (often ancestral) rather than nonhuman.

Although Mongolian shamans predominantly invoke spirits (*onggod*) from the human realm, valued precisely for their proximity to rather than their distance from human life (Swancutt 2008), they do occasionally master spirits that come from unknown clans or the nonhuman world. Many of these are regarded as 'nameless demons' (Humphrey and Onon 1996) or 'vampiric imps' (Swancutt 2008), capable of much evil-doing. Other nonhuman *onggod* are sought for their valuable insights and approached with much care and deference. Regarded as particularly powerful, wolf *onggod* belong to this latter category (Humphrey and Onon 1996: 348). Indeed, wolf *onggod* are among the few nonhuman spirits that are said to be capable of human speech (Czaplicka 1914: 231). Positioned between the human and nonhuman realms, wolves are intimately aware of human happenings and behaviours, perhaps not unlike close ancestral spirits. Capable of traversing otherwise separate domains, wolves transcend boundaries and collapse distinctions that are often seen to characterise everyday life in Mongolia.

This transcendence of boundaries between human and nonhuman realms is thus not something new and unprecedented, simply brought about by the advent of a gold rush. There is a long-standing tradition of shamanic practices that enable communication between animal *onggod*, such as wolves, and humans. Whereas the flexibility and movement of wolves into various domains is not unique to the gold mines, their unwanted appearance in human bodies outside of the shamanic trance is approached locally with much fear and hostility. Concealed in human forms, wolf people bring into question the relationship between animality and humanity. Given the remarkable agency and transformational abilities of wolves, it is perhaps not surprising that it is precisely this predator who has now entered the human domain in Uyanga. In the face of ambiguity and uncertainty, human personhood has thus become a matter of persuasion.

Rather than attempting to delineate an encompassing modality for reckoning cryptozoological species, this chapter has sought to move away from a human-centred perspective and instead take seriously the unique position of wolves in Mongolian cosmology and sociality. By recognising the continuities and ambiguities of human and nonhuman relations, we are better able to understand local people's critical comments about the Mongolian gold rush. Moreover, we are also in a position to pursue dynamic and contextual notions of morality and personhood. Rather than approaching such notions as exclusive and static human properties, we can see how they are precariously positioned within much broader cosmopolitics involving humans, spirits and, not least, wolves.

Notes

1 *Wolf Totem* is said to be serialised on radio, recast as a children's book, rewritten as a comic strip and scripted for a feature film production (Morrison 2008).
2 *The Secret History of the Mongols* is an epic of the early Mongol Empire written in the thirteenth century.
3 The Persian chronicler Rashid Al-Din (1247–1318) describes how the Mongols descended from an honourable chief called Börte Chino (see Boyle 1971). This humanised view of Börte Chino is also evident in the seventeenth-century Mongolian chronicle *Altan Tobchi* (see Bawden 1955). Written at a time of Buddhist conversion and Tibetan influence, Börte Chino is described as the youngest son of a descendant of a Tibetan ruler. For examples of Börte Chino as a mythological being, see Sinor (1982: 241).
4 There are strikingly similar stories in other parts of Asia and Europe. Perhaps the most famous story concerns the mythological founding of Rome where the twins Romulus and Remus were suckled by a she-wolf.
5 This identification of Sirius with the wolf is not unique to the Mongolian cultural region. Indeed, according to Brosch (2008: 32), the star is cross-culturally often identified with a dog, jackal or wolf.
6 For further details on conceptualisations of the *lus*, see High (2008: ch. 5) on the landscapes of spirits and High (2013a) on rumours about the emergence of new and even more powerful 'black' spirits in the mines.
7 Showing respect through honorific terms also extends to prominent mountains and certain other powerful animals (Humphrey and Onon 1996: 91).
8 Between 1926 and 1985 wolf killings in Mongolia averaged 5,308 animals annually with a peak of 18,000 animals in 1933. These numbers include only state-recognised hunts and do not take into account unofficial hunts carried out to protect livestock. As the wolf population declined rapidly, a ban was introduced between 1976 and 1980 (Wingard and Zahler 2006: 98).
9 I use the rounded average exchange rate 1 USD = 1,200 MNT.
10 In Ulaanbaatar, young men also express much interest and pride in killing wolves, and Mongolia has become a popular destination for international travellers keen on wolf hunting.
11 Although people in Uyanga complain about the selfish and ruthless behaviour of preying wolves, they do not stress a common 'humanity' among people and animals. In contrast to perspectivist cosmologies (Viveiros de Castro 1998), I have never heard informants posit that humans and animals, with their different bodies, share a unitary form of subjectivity (see also Pedersen et al. 2007: 149).
12 The term 'patriclan' here refers to a person's own nuclear family and their agnatic kindred, as well as the few previous patrilineal generations that people usually remember (rarely more than three generations).

13 Humphrey with Onon (1996: 99–100) discuss an interesting account of a person whose personality reveals the character of a wolf. Among the Daur Mongols of Inner Mongolia such animacy is regarded as a manifestation of traces of the person's soul from a previous life. Although understandings such as reincarnation are widespread in Uyanga, I have never heard informants view wolf people in this perspective.

14 The close relationship between wolves and humans is also evident in Mongolian hunting practices. In a discussion of the notion of *hiimori* ('windhorse', often translated as a principle of personally related luck), Charlier (2015) describes how the wolf is the only animal that is *hiimoritoi* ('with luck') like humans and capable of passing it on to the hunter.

References

Baring-Gould, S. 2009 [1865]. *The Book of Werewolves*. New York: Cosimo.

Baumann, B. 2008. *Divine Knowledge: Buddhist Mathematics According to the Anonymous Mongolian Manuals of Astrology and Divination*. Leiden: Brill.

Bawden, C. 1955. *The Mongol Chronicle Altan Tobci: Text, Translation and Critical Notes*. Wiesbaden, Germany: Otto Harrassowitz.

Boyle, J.A. 1971. *The Successors of Genghis Khan/Rashid al-Din*. London: Columbia University Press.

Brosch, N. 2008. *Sirius Matters*. New York: Springer Science and Business Media.

Charlier, B. 2015. *Faces of the Wolf: Managing the Human, Non-human Boundary in Mongolia*. Leiden: Brill.

Cleaves, F. 1982. *Secret History of the Mongols*. Vol. 1. Cambridge, MA: Harvard University Press.

Czaplicka, M.A. 1914. *Aboriginal Siberia: A Study in Social Anthropology*. Oxford: Oxford University Press.

High, M. 2008. Dangerous Fortunes: Wealth and Patriarchy in the Mongolian Informal Gold Mining Economy. Ph.D. thesis, Department of Social Anthropology, University of Cambridge.

High, M. 2013a. Cosmologies of Freedom and Buddhist Self-Transformation in the Mongolian Gold Rush. *Journal of the Royal Anthropological Institute*, 19(4), 753–770.

High, M. 2013b. Polluted Money, Polluted Wealth: Emerging Regimes of Value in the Mongolian Gold Rush. *American Ethnologist*, 40(4), 676–688.

Humphrey, C. 1995. Chiefly and Shamanist Landscapes in Mongolia. In *The Anthropology of Landscape: Perspectives on Place and Space*, edited by E. Hirsch and M. O'Hanlon. Oxford: Clarendon Press, 135–162.

Humphrey, C. and Onon, U. 1996. *Shamans and Elders: Experience, Knowledge, and Power among the Daur Mongols*. Oxford: Clarendon Press.

Jila, N. 2006. Myths and Traditional Beliefs about the Wolf and the Crow in Central Asia: Examples from the Turkic Wu-Sun and the Mongols. *Asian Folklore Studies*, 65(2), 161–177.

Kaplonski, C. 2004. *Truth, History and Politics in Mongolia: The Memory of Heroes*. London: Routledge.

Knight, J. 2003. *Waiting for Wolves in Japan: An Anthropological Study of People–Wildlife Relations*. Oxford: Oxford University Press.

Lindquist, G. 2000. The Wolf, the Saami and the Urban Shaman. In *Natural Enemies: People–Wildlife Conflicts in Anthropological Perspective*, edited by J. Knight. London: Routledge, 170–188.

Mishra, P. 2008. Review: *Call of the Wild. New York Times,* 4 May (accessed on 21 October 2015 from www.nytimes.com/2008/05/04/books/review/Mishra-t.html?pa gewanted=all&_r=0).

Moore, R.S. 1994. Metaphors of Encroachment: Hunting for Wolves on a Central Greek Mountain. *Anthropological Quarterly,* 67(2), 81–88.

Morrison, D. 2008. Review: *Wolf Totem. Financial Times,* 15 March (accessed on 21 October 2015 from www.ft.com/cms/s/0/afc7e59c-ee43-11dc-a5c1-0000779fd2ac.html).

Nadasdy, P. 2008. The Gift in the Animal: The Ontology of Hunting and Human-Animal Sociality. *American Ethnologist,* 34(1), 25–43.

Onon, U. 1990. *The History and the Life of Chinggis Khan (The Secret History of the Mongols).* Leiden: E.J. Brill.

Pedersen, M. 2001. Totemism, Animism and North Asian Indigenous Ontology. *Journal of the Royal Anthropological Institute* (n.s.) 7(3), 411–427.

Pedersen, M., Empson, R. and Humphrey, C. 2007. Editorial Introduction: Inner Asian Perspectivisms. *Inner Asia,* 9(2), 141–152.

Sinor, D. 1982. The Legendary Origin of the Türks. *Folklorica: Festschrift for Felix J,* 223–253.

Swancutt, K. 2008. The Undead Genealogy: Omnipresence, Spirit Perspectives, and a Case of Mongolian Vampirism. *Journal of the Royal Anthropological Institute* (n.s.) 14(4), 843–864.

Vilaça, A. 2002. Making Kin out of Others in Amazonia. *Journal of the Royal Anthropological Institute* (n.s.) 8(2), 347–365.

Vitebsky, P. 2006. *Reindeer People: Living with Animals and Spirits in Siberia.* London: HarperCollins.

Viveiros de Castro, E. 1998. Cosmological Deixis and Amerindian Perspectivism. *Journal of the Royal Anthropological Institute* (n.s.) 4, 469–488.

Willerslev, R. 2007. *Soul Hunters: Hunting, Animism and Personhood Among the Siberian Yukaghirs.* London: University of California Press.

Wingard, J.R. and Zahler, P. 2006. Silent Steppe: The Illegal Wildlife Trade Crisis in Mongolia. Discussion Papers, East Asia and Pacific Environment and Social Development Department. Washington, DC: World Bank.

6 Enigmatic bush dwarfs of West Africa

The case of the *siyawesi* of northwestern Benin

Sharon Merz

One day a hunter went hunting in the bush. Seeing that a storm was moving in, he sought shelter under a tree next to someone's homestead. He did not know that it was the home of *Kɛyawedikɛ* [singular of *siyawesi*].

The wind picked up and a branch fell from the tree. Hearing the crash, *Kɛyawedikɛ*'s dog started to bark. *Kɛyawedikɛ* sent his child to see what was happening. The child returned and told his father that there was a hunter under the tree sheltering from the rain. *Kɛyawedikɛ* sent the child to bring the hunter in from the wind and rain, was it right that his family had shelter whilst the hunter stayed outside and suffered?

The hunter entered *Kɛyawedikɛ*'s home and they greeted each other. *Kɛyawedikɛ*'s wife prepared sorghum porridge for the hunter. The hunter looked at the porridge and exclaimed, 'What's this? I can't eat this! I'm a hunter and live off the blood and meat of the animals I kill.' *Kɛyawedikɛ* replied, 'Try it and see.' The hunter started to eat and *Kɛyawedikɛ* asked, 'How do you like it?' The hunter replied, 'It is good!'

(Interview by J. Merz, February 2006)

Introduction[1]

Yammu[2], a village elder, went on to tell how *Kɛyawedikɛ* gave the hunter gifts of sorghum seed, farming utensils and poultry, together with instructions in cultivation, beer brewing and animal husbandry. The hunter, who was a member of the Benammoutchaabe – one of the twenty-three Bebelibe[3] communities – returned home and successfully followed *Kɛyawedikɛ*'s instructions (see also Huber 1979: 40–48). Benammoutchaabe community members are also known as the *siyawesi yanbɛ* (*siyawesi* 'owners'). Today, this story of the hunter's encounter with the *siyawesi* bush dwarfs is recounted when explaining how the Bebelibe discovered sorghum, yam, beans and other foods (see also Huber 1973: 386).

With a population of around 69,500 people, the Bebelibe are largely rural and live in loose-knit villages in the mountainous area of the Atacora region, in the northwest of the Republic of Benin. Most of the Bebelibe are located in the Commune of Cobly, with additional villages in the neighbouring Commune of Boukoumbé and across the border in Togo. Many rely on subsistence farming and growing cash crops for their livelihood. All of the Bebelibe with whom I have spoken testify that *Uwienu* is the Supreme Being and creator of all. The first Catholic missionaries arrived in the Commune in the late 1940s, whereas the first Protestant missionaries arrived in the early 1950s (Akibo 1998; Cornevin 1981: 436, 440–441, 453–454). Whilst an estimated 10 per cent of the population now attend a church, Islam remains marginal. The majority of the population continue to be guided by the beliefs and practices of their ancestors and for many people paying homage to *Uwienu* through his intermediaries is an integral part of life. These intermediaries include community/lineage guardian shrine entities, ancestors and the *siyawesi* (see also Huber 1973: 378). Since their initial encounter with the hunter, the *siyawesi* have maintained an alliance with the Bebelibe and regularly choose individuals to become diviners.

In this chapter I present the case of the *siyawesi* bush dwarfs and their ongoing relationship with the Bebelibe. Such human-like beings are known throughout the region and hold positions of varying importance in West African ontology. Bush dwarfs are frequently referred to in ethnographic works dating from the 1920s onwards (e.g. Cardinall 1927; 1931; Rattray 1927; Jackson 1977a; 1977b; Fainzang 1986; Kirby 1986; Maurice 1986; Erny 1988; Ovesen 1990; Tengan 1990; Bonnet 1981–82; 1986; 1994; 1995; Goody 1997; Guigbile 2001; Koabike 2003 and Heraud 2005) often in subsections dealing with ontology, divination, sacrifice, aetiological and procreation beliefs or social structure. To date, the only published literature that mentions the *siyawesi* specifically has been written by Hugo Huber (1973), a Swiss anthropologist who conducted research amongst the Bebelibe in 1966–1967.

Although people across the region usually refer to these entities as bush dwarfs or bush spirits, different authors have employed a wide range of vocabulary when interpreting the given vernacular term, which suggests that there is some uncertainty about how these entities should be viewed and classified. Maurice (1986: 448–450) alone refers to them as *nains* ('dwarfs'), *démons* ('demons') and *esprits* ('spirits'). They have also been labelled as 'fairies' (Bannerman-Richter 1987: 30; Goody 1997: 64 n.69; Rattray 1927: 25), 'pixies' (Cardinall 1931: 77–96) and 'sprites' (Goody 1962: 212). I prefer to call them 'bush dwarfs' as terms such as 'spirit' do not adequately express their embodiment for those who see and relate to them, nor does it convey their lifestyle, which reflects – and is thought to precede – that of their human counterparts, as the opening story demonstrates (see also Goody 1997: 65–66; Kirby 1986: 61–62; Ovesen 1990: 150; Tengan 1990: 6–7). I recognise that the term 'dwarf' also has its limitations, as it runs the risk that they are reduced to mythological beings. Thus, when I write about the specific situation for the Bebelibe, I employ the Mbelime terms *kɛyawedikɛ* and *siyawesi*.

I consider the case of the *siyewesi* pertinent. Although the many descriptions prevalent in the existing literature indicate an underlying assumption that – for the societies concerned – these entities truly exist, little has been written that addresses the reality of their existence. One exception is

Bannerman-Richter's (1987) autobiographical account about the *mmoetia* of Ghana. Bannerman-Richter recounts his journey from scepticism to accepting that the *mmoetia* exist after personally meeting and interviewing some of them in 1970. Having lived amongst the Bebelibe for over thirteen years, I cannot dismiss the *siyawesi* as imaginary (see also Geschiere 1997: 20–21). I am in the same position as Geschiere (1997: 23) was when he wrote about witchcraft: although I cannot personally verify the *siyawesi*'s presence, I cannot deny the obvious reality of the *siyawesi* for those I have spoken with. I hope to demonstrate that the question of the *siyawesi*'s reality should be taken seriously, whilst also showing that how people engage with them changes depending on different trajectories of modernity (Geschiere et al. 2008: 5).

I start this chapter by describing the *siyawesi*'s appearance and behaviour before moving on to examine their place in Bebelibe ontology and role in divination. With reference to testimonies given by those I interviewed, I subsequently address the question of reality and the impact of post-colonial modernity as represented by the arrival of institutions such as churches, Western-style health care and an education system modelled on the French one. I conclude by proposing that bush dwarfs are not merely mythical beings but that they really exist.

Siyawesi appearance and behaviour

A major factor differentiating *benitibe* ('people') from *tiwante* ('animals') is their ability to communicate directly.[4] The *siyawesi*, however, are not considered animals. Although they are invisible to most humans, their bodily form resembles that of a human and they speak directly with those they choose to relate to. They know all languages and will speak the language of the person concerned. Those I interviewed indicated that the *siyawesi* are between half and one metre tall and have long, thick hair. Consequently, they are also known as the *tiyuute yanbɛ* ('hairy ones'). They live in the bush, usually go around in pairs, cultivate red rice and sorghum and rear black-coloured helmeted guineafowl (*Numidameleagris*). They like to play in water and often come out at night when they sow in people's fields and either take or replace grain from people's granaries (see also Huber 1973: 386–387). In more recent years, this behaviour has expanded beyond the granary to include money and other valuable items.

Everybody I talked to shared similar accounts about the nature of the relationship between the *siyawesi* and people. If you are chosen by the *siyawesi* and you maintain a good relationship with them, you will do well and become rich. They will provide you with heightened perception, see that your crops and domestic animals succeed, make sure you never lack for simple provisions and generally take care of you. If you do not maintain the relationship, then you will suffer. The *siyawesi* can hide things from you, chase you and whip you, steal your provisions and give them to those in their favour and see that you fail in your endeavours (for a similar description see Kirby 1986: 61–62, who writes about the Anufɔ of Ghana and the *jinn*. Cf. High, this volume). Esther, speaking from personal experience, explained that the '*siyawesi* can give life but can also bring suffering [...] You see, it's like this, what they do is harsh and yet is good [...] if you can provide for them, you will be rich [...]' (Interview, October 2009).

All the interviewees gave similar explanations about what you need to do to keep the *siyawesi* happy: give them the first fruits of your harvest, the firstborn of your domestic animals and any provisions they may ask for. Jackson (1977b: 126), who wrote about sacrifice and the social structure of the Kuranko of Sierra Leone, explained that the Kuranko offer sacrifices to bush spirits in recognition of their allowing the Kuranko to settle and farm the land and that such offerings are given indirectly to God. Those who offer gifts to the *siyawesi* do so with the same understanding. The *siyawesi* particularly appreciate tiger nuts (*Cyperusesculentus*), bambara groundnuts (*Vignasubterranea*), peanuts, sesame seeds and non-fermented sorghum beer (see also Huber 1973: 386–387). After Esther's father died, providing for the *siyawesi* became a challenge. Unable to satisfy their demands, they started to trouble her. In the end she took refuge in a church: 'I asked *Uwienu* (God) and he sent the *siyawesi* away from me, as I don't have someone who can perform sacrifices for me.' (Interview, October 2009).

Interviewees explained that the *siyawesi* usually choose people when they are young. Such people wander off into the bush and cannot be found, sometimes for several days or even weeks at a time. They turn up later well fed and taken care of. They may be troubled with ideas and thoughts, see things that other people miss and know things about others without knowing where the knowledge came from. Robert added that the *siyawesi* befriend the children they choose and that children usually go willingly with the *siyawesi* into the bush. He then spoke about his younger brother who regularly disappeared into the bush with the *siyawesi*. Several of those I interviewed shared that these are signs that a child has been chosen to become a diviner. Robert's brother would have become a diviner had he not died. Kodani explained that everyone chosen by the *siyawesi* could become a diviner, but not everyone will pursue this end. He clarified that most women do not become diviners as the diviner sits in a way that makes it relatively easy for male clients to look up the women's skirts.

The relationship between bush dwarfs and diviners is not unknown in West Africa. Bonnet (1986: 20–21) describes a similar situation for the Mossi of Burkina Faso. The vocation to become a diviner, who is usually a woman, typically starts with '*un épisode psycho-pathologique*' [a psycho-pathological episode] (1986: 21) during which the would-be diviner enters into direct communication with the *kinkirse* ('bush spirits') and disappears into the bush for several weeks. The time spent in the bush is understood as a period of initiation when the *kinkirse* share their knowledge of plants and trees with the would-be diviner. Once the person returns home, her family give food known to be appreciated by the *kinkirse* and the newly appointed diviner begins an apprenticeship under another experienced diviner.

Maurice (1986) provides a detailed ethnography of the Betammaribe, southeastern neighbours of the Bebelibe. In a chapter dealing with magic Maurice (1986: 448–450) briefly describes the *dékiaribidé*, whose appearance and behaviour closely resemble the *siyawesi*. He mentions that the *dékiaribidé* seek relationships with seers (*inintînwân* 'those with eyes') who then consult the *dékiaribidé* when someone is sick. Whilst Goody (1997: 65–66), writing about the LoDagaa of Ghana, mentions that the *kontome* have human form, can speak, have feelings, converse with humans and that they too act as intermediaries passing on information that has been revealed to them during divination sessions. Finally, when describing Sisala ontology (Ghana), Tengan (1990: 6–7) includes a brief description of *kantomo* 'spirits of the wild' who are 'the 'first borns' [sic] of Wiise [Supreme Being] and his messengers to humans' (1990: 6).

I now examine the *siyawesi*'s role in divination and their place in Bebelibe ontology in more detail.

Siyawesi's role in divination and place in Bebelibe ontology

Huber notes that there are the two principal groups of intermediaries between the Bebelibe and *Uwienu*: the community/lineage guardian shrine entities and the ancestors. He then mentions the *siyawesi* as a third category and states that; '*Une certaine importance, plutôt mythique que rituelle, est également attribuée aux* siawissi, *petits génies de la brousse.*' [A certain importance, more mythical than ritual, is attributed to the *siyawesi*, small spirits of the bush.] (1973: 378).

Huber describes what happens should someone encounter the *siyawesi*:

Possédé, il court dans la brousse comme un fou; il commence à crier et on court après lui pour l'attraper; on le conduit chez le devin, qui est le gardien de la médicine des siawissi. La possession par ces génies est, en effet, jugée comme un signe de la vocation de devin.

[Possessed, he runs into the bush like a madman; he starts to shout and others run after him to catch him; he is taken to the diviner, who is the guardian of the siyawesi medicine. Possession by these spirits is, in fact, judged as a sign of vocation as a diviner.]

(1973: 387)

Huber then explains that a dwelling place for the *siyawesi* needs to be installed in the home of the person possessed. This is usually in the form of a statue with a hole in its stomach to receive offerings. A diviner then conducts a dedication ceremony that includes pouring libations and animal sacrifices (1973: 387).

Huber (1973: 388) makes further reference to the link between the *siyawesi* and the diviner before going on to describe the importance of the diviner for the Bebelibe with the expression '*Les choses du devin sont la vie de l'homme.*' [The things of the diviner are the life of man.] (1973: 389). The diviner is consulted in all circumstances: during sickness, infertility, pregnancy and droughts, after accidents, following births, deaths, strange dreams, and so on (1973: 389). Huber then asserts that Bebelibe thought '*élève donc la divination clairement au-dessus d'une simple technique ou d'un simple jeu magique*' [rates therefore divination clearly above simple technique or a simple magical game] (1973: 389) and that '"*Aller chez* Uyenu" *et* "Uyenu *a dit*" *sont les synonymes de* "aller chez le devin" *et* "le devin a révélé". *Cela n'exclut pas la croyance générale selon laquelle le charisme professionnel du devin serait inspiré par les* siawissi, *les génies de la brousse*.' ['Going to *Uwienu* (God)' or '*Uwienu* said' is synonymous with 'going to the diviner' and 'the diviner revealed.' This does not exclude the general belief that the professional charisma of the diviner is inspired by the *siyawesi*, the spirits of the bush.] (1973: 430).

The fact that Huber noted that people generally believe that diviners are inspired by the *siyawesi* (1973: 430), whilst also relegating their role as *mythique* ('mythical') (1973: 378), suggests that he is sceptical about the *siyawesi*. My research affirms that the *siyawesi* not only inspire the diviners, but that their intermediary role is essential for divination to take place and cannot be dismissed as mythical.

The root of the word *kɛyawedikɛ* is made up from the verb *ya* ('to know or see') and the word *wedi*, which is employed when describing how toddlers walk, thus it literally means 'the waddling seer'.

Those chosen by the *siyawesi*, whether they pursue a vocation as a diviner or not, are usually members of the Benammoutchaabe community or have a matrilateral family link with the community. In addition, the chosen individuals have the *mtakimɛ* of divination. The *mtakimɛ*, put simply, can be translated as 'destiny' (see also Huber 1973: 384; Merz 2014: 93). After a time, a new relationship with the *siyawesi* needs to be formalised. Failure to do so can result in the *siyawesi* disciplining the individuals concerned by whipping and chasing them. Individuals may appear to be mad as they shout out, dance around or run off and hide to avoid being whipped. Once the family formalises the relationship, this apparent madness will disappear and only returns should somebody attempt to sever the relationship and thus incur further discipline from the *siyawesi*. Esther described from her personal experience how this could appear to an onlooker:

> The *siyawesi* will be whipping me. You will see me crying but you won't see why. I will flee and hide under that car [points to vehicle] or in there [points to chicken coop]. If they were to say, 'take your clothes off and throw them down,' I would do it. It can drive you crazy. Don't you see? If you are able to arrange the relationship, it will work well and it is good.
>
> (Interview, October 2009)

Basadi, a diviner, and Kodani both affirmed that the *siyawesi* madness is different from normal mental disorders and will stop as soon as the relationship has been formalised.

When Huber (1973: 387, see above) described similar behaviour resulting from an individual's encounter with the *siyawesi*, he employed the words *possédé* ('possessed') and *possession*. I consider his interpretation misleading, as the *siyawesi* never possess individuals through invasive or 'executive' possession (Cohen 2008). During my interviews, in both Mbelime and French, interviewees used the verbs 'follow', 'choose', 'catch' or 'whip'.

Koabike (2003) who writes about his own people, the Moba of northern Togo, includes a description of the *sanpola* (2003: 44–46). He notes that the *sanpola* are short beings with long hair and human form and that the Moba also refer to them as *niib* ('humans'). Koabike (2003: 44) calls them *démons* ('demons'), as the *sanpola*[5] are considered malevolent. During a personal conversation (15 July 2011), Koabike explained that the Moba sometimes employ the verb 'to possess' when describing the *sanpola*'s actions in French. He went on to explain that this was for want of a better word. He chuckled at the idea of the *sanpola* actually possessing someone: 'How would they fit into the person's body?' For Koabike the outcome is the same as being invasively possessed, as the *sanpola* dominate the person and force the individual to do their bidding. Cohen (2008) refers to this as 'pathogenic possession'. Although the *siyawesi* are capricious and can be forceful, as Esther's example above demonstrates, they are not generally considered malevolent. This is reserved for *Disɛnpode*[6] and the *sihonkpaasi*. I discuss these beings in more detail below.

Those chosen by the *siyawesi* have two small clay statues (male and female) made for them. The installation of the *siyawesi* statues is crucial for the well-being of the individual concerned as it allows for better communication between the individual and the *siyawesi* and for the chosen person to be at peace. Kodani, who is from the Benammoutchaabe community and has a relationship with the *siyawesi*, regularly constructs *siyawesi* statues and performs their installation for people. He describes how he came to construct statues as follows:

> The *siyawesi* caught me when I was young. My parents had the *siyawesi* constructed for me. The *siyawesi* then caught my children and I constructed *siyawesi* for them […] So I said to myself, 'ah, is this your path? You must make and follow the path. If *kɛyawedikɛ* catches someone, you must go to him and make the *siyawesi*.' That's why I construct *siyawesi* for people. How do I know how to construct *siyawesi*? It is the *siyawesi* who show me and tell me, 'put this here, do that, take this […]' It is *Uwienu* [God] who sends *kɛyawedikɛ* and says, 'go to that person and become friends.' If *Uwienu* does not send the *siyawesi*, they don't go.
>
> (Interview, June 2011)

Kodani's account is important as it shows that he does not distinguish the *siyawesi* from their statues – they are both '*siyawesi*'. In order to comprehend the role of the *siyawesi* statues, a general understanding is needed of the

Bebelibe's underlying ontology, which corresponds with what Ingold (2000: 112–113) describes as an animic ontology (see also Merz 2014: 79–135). Bebelibe engagement with the world is intersubjective or relational (see also Bird-David 2006: 44; Hornborg 2006: 29; Kohn 2007: 4). Following Bird-David (2006: 47–48), such a relational view in turn creates a shared community. Bird-David (2006: 47–48) further develops her argument for relationality by pointing out that depicting the other necessarily involves taking an objective perspective, thus the visual representation of animals is often lacking in such societies. The same situation exists for the Bebelibe who normally do not represent other beings by any visual means (see also Merz 2015: 104–105).[7] The *siyawesi* – who are classed as human – are the exception. This reluctance to represent the other is present among other peoples of the West African savannah as demonstrated by Goody, who notes the same thing for the *kontome* bush dwarfs and LoDagaa of Ghana: 'Virtually alone among the LoDagaa, [the *kontome*] are given human images in sculpted form. Figurative representation, 'breath' and resemblance to mankind are closely intertwined' (1997: 66).

These statues, however, should not be considered representations of the *siyawesi* as the statues are seen as identical with the *siyawesi*. Rather, they help orientate the relationship between the person and the *siyawesi* (see also Merz 2014: 79–135), which is why the statues appear to be crude to outsiders. Following Ingold (2000: 130), such statues are important not for what they depict, but for what they reveal, in this case the existence of a relationship with the *siyawesi*. Ingold continues by stating that the purpose of such depictions is 'to penetrate beneath the surface of things so as to reach deeper levels of knowledge and understanding. It is at these levels that meaning is to be found' (2000: 130). This being the case, it does not matter if the statues get damaged.

For Kodani, the installation of the *siyawesi* equals becoming a diviner. Kodani went on to describe the *siyawesi* installation:

> You see, you need [ten things such as] sesame, salt, eel, catfish, bush meat, beef, [local] cheese, tiger nuts, fresh peanuts, bambara groundnuts – that makes ten things. Do you see? You need to have all ten things. Now you need to find a black chicken, a black [helmeted] guineafowl together with a black goat. Oh yes, you need a black dog too. Many things are needed. [Once everything is ready] I come with the *siyawesi*. There is also non-fermented and fermented sorghum beer [...] Once I have arrived the men of the family give me rice flour. [Having mixed some rice flour with non-fermented sorghum beer] I pour libation [on the *siyawesi*]. I then sacrifice the animals and I install the *siyawesi*. I give the person his divining staff and wash his face with blood [from the sacrificed animals] [...]
>
> (Interview, June 2011)

The installation includes drinking *siyawesi* medicine made from special tubers that resemble yam and grow wild in the bush. This medicine, together with washing the person's face, further heightens the individual's perception. Both Kodani and Basadi told me that once the person receives the diviner's staff, she/he immediately starts to divine.[8] Basadi added that a person only drinks the *siyawesi* medicine during the *siyawesi*'s installation; it is not something that the person drinks regularly. Once the installation has been completed, the relationship between the person and the *siyawesi* is now formally recognised.

The *siyawesi* are therefore synonymous with divination. Without the *siyawesi*, the diviner knows nothing. Tcheteka explained that the *siyawesi* speak directly to the diviner and that when someone comes for a consulting session, 'the diviner takes his staff and sits and the *siyawesi* come. They tell him to say, "perform this sacrifice; do that; tell him [the one seeking advice] to sacrifice a cow, sacrifice a dog; tell him to prepare sorghum beer". That's how it happens [...]' (Interview, November 2009).

Esther shared how the *siyawesi* would reveal things to her through dreams whilst she slept or through placing special knowledge in her head. She reiterated that it is the *siyawesi* who are really the diviners. The knowledge they place in her head pushes her to act, she cannot ignore it: 'It happens in my ideas, telling me to talk. If I refuse by saying, "I can't", the thoughts will plague me [until I talk]' (Interview, October 2009).

Antoine compared a divination session with our interview. Diviners ask questions on behalf of the client and the *siyawesi* tell them what to say. Antoine explained that a diviner stands up to signal the end of a divination session as the *siyawesi* have left.

All those I spoke to affirmed that *Uwienu* (God) created the *siyawesi*. Several then explained how the *siyawesi* know what to tell the diviner. For Kodani and Basadi, the *siyawesi* are *Uwienu*'s messengers who communicate with the diviner on *Uwienu*'s behalf and thus mediate between *Uwienu* and the Bebelibe:

It's *Uwienu* who gives the *siyawesi* knowledge, which they follow. They come and tell you, 'this is how it is, *Uwienu* said that this is how it is'. If you want to do something, you first call on *Uwienu* don't you? Yes, and it's the *siyawesi* who answer. It's *Uwienu* who has the *siyawesi*; he created them. Yes, the *siyawesi* are *Uwienu*'s messengers. It's *Uwienu* who gave them their knowledge. *Uwienu* created man and he created *kɛyawedikɛ*.

(Interview with Kodani, June 2011)

The *siyawesi* live in the bush. When the diviner arrives, he takes his staff to call the *siyawesi* [Basadi taps his staff on the ground]. The *siyawesi* arrive and sit there [points to the ground beside him] to interpret for you (the diviner). Then you explain to the person (the client) [...] It's *Uwienu* who gives the message and the *siyawesi* pass the message on to us. It's *Uwienu* who tells the *siyawesi* to do this or that. It's *Uwienu* who created them.

(Interview with Basadi, June 2011)

Esther, who has taken Jesus in place of the *siyawesi*, still believes that they are *Uwienu*'s messengers: '*Uwienu* brought the *siyawesi* forth and it is *Uwienu* who tells them what to say. *Uwienu* brought them forth and they are *siyawesi*. It is *Uwienu* who brought the diviner forth. People say that if you have *siyawesi*, you are a diviner' (Interview, October 2009).

Antoine, who used to attend the Catholic Church, has a different perspective. He believes that the *siyawesi* follow commands issued by *Disɛnpode*, who is now commonly equated with the devil (see below). He explained that it is *Disɛnpode* who sends sickness. When a child becomes ill, the father goes to consult a diviner to find out why his child is sick and what he should do about it. Meanwhile, *Disɛnpode* sends the *siyawesi* to tell the diviner what needs to be sacrificed for the child to get better. For Antoine this is all a ploy to get food for *Disɛnpode* who is fed by the sacrifices: 'Your child is sick, you take a cow, you take a goat, you take a chicken to perform sacrifices and it's *Disɛnpode* who eats' (Interview, November 2009).

Tcheteka, a Protestant church elder, believes that the *siyawesi* are *Disɛnpode*'s children. He explained that *Disɛnpode* listens in to *Uwienu*'s conversations. *Disɛnpode* then uses this knowledge to his advantage, passing it along to the *siyawesi* who then share this knowledge with the diviner: 'when *Uwienu* talks, *Disɛnpode* listens. He knows the good that *Uwienu* wants to do. Don't you see? *Siyawesi* are *Disɛnpode*'s children and he sends them everywhere' (Interview, November 2009).

As Tcheteka and Antoine talked about *Disɛnpode* and the *siyawesi*, it was evident that the teaching they have received in church has influenced their ontology and how they now perceive these entities. They explained how *Uwienu* first created *Disɛnpode* and the *siyawesi*. They then disobeyed *Uwienu* who sent them away to live in the bush. Following the estrangement in their relationship, *Uwienu* created humans. *Disɛnpode* became jealous of *Uwienu*'s new rapport with humans and set out to destroy this bond. Tcheteka went on to retell the Genesis account of how *Disɛnpode* tricked Adam into sinning.

Disɛnpode and *sihonkpaasi*

In order to better understand some of the changes in how the *siyawesi* are perceived, I now briefly examine two other enigmatic beings: *Disɛnpode*, who also inhabits the bush, and the *sihonkpaasi* who are associated with rivers and the sky. Changing beliefs about these entities have, in turn, impacted people's opinion of the *siyawesi*. Whilst *Disɛnpode* is growing in importance, knowledge about the *sihonkpaasi* is diminishing.

Descriptions are a lot more ambiguous for *Disɛnpode* compared to the *siyawesi*. Nenboni, a village chief, described *Disɛnpode* as follows:

> *Disɛnpode* lives in the bush. He doesn't go into people's houses as the *siyawesi* do. He has thick hair and a long neck and if he sees a human, he flees. Those who are strong if they see *Disɛnpode* they will hunt him down

and take some of his hair. This is why some people have hair like the whites. That's *Disɛnpode. Kɛyawedikɛ* though is human.

(Interview by J. Merz, July 1997)

Whilst Yammu depicts *Disɛnpode* as

an evil being who lives in the bush. If he touches or catches someone, the person dies. *Kɛyawedikɛ* and *Disɛnpode* are not the same. *Disɛnpode* comes from the bush. *Uwienu* created him and he lives in the bush. If *Uwienu* allowed him to enter people's homes, it would be the end. *Uwienu* chased him from human habitation into the bush because he is evil. *Disɛnpode* is nasty.

(Interview by J. Merz, February 2006)

Esther mentioned *Disɛnpode* in passing as a malevolent bush being who does not enter people's houses, whilst Kodani explained that *Disɛnpode* is bad whereas the *siyawesi* are good. Huber (1973) made no mention of *Disɛnpode*, which suggests that his role in Bebelibe ontology was insignificant during the 1960s. Today, *Disɛnpode*'s place in Bebelibe ontology has been enhanced since the first missionaries chose the term *Disɛnpode* for the devil, although descriptions about him remain somewhat fuzzy. Everyone agrees, however, that *Disɛnpode* is evil. J. Merz (2008: 208) explains that most Bebelibe have accepted *Disɛnpode* as the devil regardless of their faith, which has further obscured pre-colonial beliefs about him. As it seems that *Disɛnpode*'s pre-Christian role in Bebelibe ontology was not well established, this may have facilitated his adoption as the devil.

Sihonkpaasi literally means 'river beings who paralyse'. They are also known as the *ihiini yanbɛ* ('sky owners'). Several people affirmed that they are short and human-like in appearance but, unlike the *siyawesi*, only have two limbs (an arm and a leg) and short hair. They like to play in water but primarily live in the sky. They can fly and are extremely rapid. Their name is a compound noun that is not originally Mbelime. Indications are that these entities originate from the Berba, northern neighbours of the Bebelibe, and that they have since gained a foothold in Bebelibe territory.

The *sihonkpaasi* are malevolent beings and do not befriend people. They are best known for attacking children and leaving them partially paralysed or mentally retarded, assuming they do not die. Their attacks are quick and severe. The children collapse, their eyes roll to the back of their heads and their limbs become stiff. It seems that to date no medical explanation has been found and those taken to the hospital usually die.

During several of my interviews, people mentioned that the *siyawesi* sometimes turn nasty and attack people with paralysis. Robert explained that many people now confuse the actions of the two beings, attributing the evil acts of the *sihonkpaasi* to the *siyawesi*. Consequently, knowledge about the *sihonkpaasi* is slowly diminishing, whilst some people now perceive the *siyawesi* more negatively than before. Several people blamed this confusion on new ideas that have been introduced by schools and churches. To understand what may be behind these changes in how the *siyawesi* are perceived, I now discuss the question of reality and the impact of post-colonial modernity.

Reality and modernity

Most people have a tale to tell either of their personal experience of the *siyawesi* or having witnessed someone else's encounter. The reality of the *siyawesi* cannot be denied for those Bebelibe who have shared their experiences with me either informally or formally these past thirteen years. Cardinall wrote that: '[t]hroughout Togoland, as in Ashanti and the Northern Territories [of Ghana], there is a confirmed belief in the existence of mischief-loving dwarfs. The Ashanti call them mmotia and in the north they are known as kulparga or chichiriga' (1931: 77).

Rattray (1927: 22–26) also has a brief description of the *mmoatia* and noted that peoples' beliefs and descriptions of the *mmoatia* are so credible that 'perfectly stolid, matter-of-fact Englishmen' (1927: 25) have been convinced of their existence. The general conviction about the reality of such entities seems to have remained steadfast despite the fact that today's West African societies have changed significantly since Rattray's time.

The Bebelibe distinguish two eras: *ubɔɔyɔ* ('old times') and *upaanu* ('new times'). The latter term is employed in a variety of senses including all that is new: things, ideas, institutions and new ways of doing things. *Upaanu* is often translated into French as *la modernité* – which can be equated with post-colonial modernity – and is used to demarcate the arrival of European colonialists and the ongoing changes that are associated with them. Institution-wise, in the Commune of Cobly the first mission station and school was established in 1947; the first health centre in 1951; the first state primary school in 1959, the state secondary school in 1985 and the 6th form in 2009.

Piot, writing about the Kabre of northern Togo, suggests that the savannah region in general 'has long been globalized and is better conceptualized as existing within modernity' (1999: 1). He realises that this may seem contrary to appearances with the Kabre's apparent 'earmarks of a still pristine African culture'. He goes on to suggest that Kabre's 'apparently traditional features [...] owe their meaning and shape' as much to their 'encounters with Europe' as to their local origins (1999: 1). The Kabre's interactions with others and appro-priation of ideas from elsewhere are deliberate and the society cannot be considered bounded or internally focused, nor is it possible to separate 'tradition' from 'modernity' (1999: 16–24, 173–174; see also Geschiere 1997: 8). Piot notes that the Kabre 'welcome and appropriate many things Western – tin roofs for their houses, Western clothes and medicine, radios and cars, a moneyed economy, certain forms of Christianity [...]' (1999: 23).

Piot's observations are equally valid for the Bebelibe. Despite local tendencies to dichotomise between those who are 'traditional' and those who are 'modern' – itself a product of modernity (Geschiere et al. 2008: 3) – all Bebelibe live and participate in *upaanu* ('new times'). How different individuals choose to interact with *upaanu* varies depending on their personal experiences and what they choose to appropriate. Following Geschiere et al. (2008: 5) different trajectories of modernity are apparent within any given society. Taking the testimonies of Kodani, Basadi, Esther, Tcheteka and Antoine, three main trajectories are apparent:

Kodani and Basadi both have personal relationships with the *siyawesi* and do not attend church. For them the *siyawesi* are *Uwienu*'s (God's) messengers; Esther, a Christian, also has a personal relationship with the *siyawesi*. Since her conversion, her interaction with them is minimal. Despite some of their seemingly capricious behaviour towards her, she has not demonised them. She has experienced their favour as well as their disfavour. They remain *Uwienu*'s messengers but have been superseded by Jesus; Tcheteka and Antoine have not had personal encounters with the *siyawesi* but have attended church. They do not think the *siyawesi* are *Uwienu*'s messengers but associate them with *Disɛnpode* ('the devil'). Despite, this Tcheteka and Antoine remained ambivalent about the *siyawesi*. Geschiere points out that such ambivalence is why 'discourses on the occult incorporate modern changes so easily' (1997: 13). None of the five have been to school. Their differing perceptions of the *siyawesi* appear to be largely affected by their relationship – or lack of – with these entities and whether they have attended church. Their differing experiences have thus modified how they perceive the *siyawesi* but they do not deny the *siyawesi*'s existence. Pels, writing about magic and modernity, observes that 'even if magic is denounced as "bad" tradition and backward, it is often, in the process, reconstructed in novel ways that may come to influence social practices at a later stage' (2003: 30). The same can be said of the *siyawesi*.

Many people are attracted to Christianity, as it allows them to move along a trajectory that offers an immediate association with post-colonial modernity, which often results in a transition towards individualism, promotion of the nuclear family, increase in ownership rights and commerce (Bayart 2008 [1998]: 92–93; Comaroff and Comaroff 1992: 200–201; Erny 2001: 19; Horton 1971: 86; Manning 1998: 101; Merz 2008: 209; Meyer 1998; 1999). This is evident amongst Bebelibe Christians, who are often the most economically active and reject many of the local customs. There are many, however, who have attended church, become disillusioned by it and have left again. This disillusionment seems to be with the institutional nature of the church rather than with Christian teaching. Thus, they have happily appropriated ideas from the teaching they received, as the example of *Disɛnpode* illustrates, especially if the ideas help them to better understand the world around them.

Other institutions associated with *upaanu* ('new times') include Western-style health care and an education system modelled on the French one. People readily accept information about viruses, bacteria and parasites as disease agents, for example, but this scientific knowledge does not address *why* an individual became sick in the first place. Only *Uwienu* knows this. After visiting the health centre, especially if the person does not recover as expected, those who have converted to Christianity intercede for the sick person through prayer, whilst others consult *Uwienu* via the diviner, who in turn consults the *siyawesi* to discover the cause of sickness. Some do both.

Jean, an undergraduate student, studied philosophy and biology as part of his *baccalauréat* (equivalent of 'A' levels). Consequently, he started to question his underlying animic ontology. His attitude to animals indicates a shift from viewing them as beings of equal worth to considering them as inferior to humans. His conviction that the *siyawesi* exist, however, stays solid. Although he has not personally encountered the *siyawesi*, he and several friends shared a homestead with someone who has, and he witnessed this person's encounters with them. He concluded that just because science cannot prove that the *siyawesi* exist, does not mean that they do not.

Bonnet (1995: 502) suggests that scientific theories do not necessarily invalidate people's beliefs. This, in turn, explains why many Africans who work in the medical profession can appear contradictory when explaining the cause of a sickness, as they alternate between their local and scientific knowledge. The example of Bannerman-Richter (1987), however, suggests that with time and prolonged exposure, such convictions can be quashed. Bannerman-Richter, who moved from Ghana to the USA as a young man, explained why his childhood beliefs in *mmoetia* turned to scepticism:

> My doubts were probably created by my constant exposure to Western ideas and values, and by my mission school education. Serious Western literature does not condone a belief in the supernatural; the Church of England (Episcopal Church), to which I belonged frowned on such beliefs and regarded them as superstitions and heathenisms […]
>
> (1987: 55)

Bannerman-Richter adds that he initially assumed that beliefs in the *mmoetia* were limited to 'mere ignoramus[es]' (1987: 71). His return to Ghana and personal encounters with the *mmoetia*, however, reversed his scepticism. Gilbert (1998: 81) notes that beliefs in the *mmoatia* are associated with the rural, uneducated poor. This Western scepticism and dismissal of bush dwarfs as being the machinations of the rural and uneducated poor, however, is not tenable amongst the Bebelibe, as demonstrated by Jean above and André, a retired pastor, who told me that he had been 'taken' by the *siyawesi* when he was young; they took him into the bush where they fed him. Now he is a Christian, he does not actually see them anymore but he knows they are present. André reiterated throughout our interview that the *siyawesi* are real. At one point he insisted, 'they exist, they exist. You can't just do away with (*supprimer*) them. They will always be there' (Interview, June 2011). A key difference between Bannerman-Richter's and André's experience lies in how the church itself views such beings. Bannerman-Richter attended a church in the USA where his beliefs were not taken seriously, whilst the local churches in Cobly do not deny that the *siyawesi* exist, but rather demonise them.

Hill, writing about witchcraft beliefs in Africa, makes the following observation:

> One limitation to the majority of the research [...] is that the scientists do not believe in the reality of the spiritual or psychic world. Discounting the supernatural, all is reduced to sociological or psychological causes. Their theories contain truth, but they are partial explanations.
>
> (1996: 325)

She suggests that one reason Westerners struggle with such beliefs is due to their lack of knowledge and not because witchcraft does not exist (cf. Bernard, this volume). She realised that her ignorance contributed to seemingly illogical data: 'Inconsistent responses lead me to doubt the logic of my questions rather than the logic of the African worldview' (1996: 334).

Van Dijk and Pels note that: 'the extent to which one privileges a "natural" over a "supernatural", or a "scientific" over a "religious" conception of the world is the outcome of a struggle over how to perceive, rather than the reflection of a given "objectivity"' (1996: 249). They then go on to discuss Western perception, based on what the eye sees of reality, versus '"inner" vision or "inspiration"'. They 'feel that a study of the politics of perception in the practice of studying religion in Africa can restore initiative and agency' to those under study (1996: 250; see also Pels 1998: 205). Such an approach resonates with much of the literature on the more recent ontological turn in anthropology (e.g. Henare et al. 2007; Scott 2013; Kohn 2015). I have attempted to do this by allowing several Bebelibe to speak for themselves in this chapter. Yammu explained to me that we do not really see with our eyes, but rather with our *kɛbodikɛ* ('vital force'): 'it's *kɛbodikɛ* that sees, it's *kɛbodikɛ* that sees. The eye sees nothing; it's *kɛbodikɛ*' (Interview, February 2012).

Thus, it is crucial that one examines people's convictions, such as those surrounding bush dwarfs, in the light of the society's own discourse and practices, rather than one embedded in Western epistemology, as J. Merz (2004: 574) suggests with regard to witchcraft beliefs. He (2004: 573) points out that colonial administrators and missionaries expected beliefs about witchcraft to disappear as a result of modernisation, as they thought had happened in Europe following the Enlightenment. This has proved not to be the case – neither in Africa nor in Europe (see also Geschiere 1997). J. Merz observes that: 'The question of the "reality" of witchcraft and other beliefs is already an old dilemma in anthropology because beliefs, by their very nature, escape the scientific scrutiny of the Enlightenment paradigm as they are intrinsically empirically unverifiable and unfalsifiable' (2004: 574).

The dilemma continues. Despite advances in science and technology, and new knowledge that is promoted through Western-style health care and French-inspired schooling, experiential certainty of the *siyawesi* persists.

Meanwhile, other influences, such as Christianity, help to maintain this certainty whilst also modifying it. Consequently, *Disɛnpode* has gained in importance in Bebelibe ontology through his transformation into the devil, whilst the *siyawesi* are in the process of becoming demonised by some and have taken on certain characteristics of the *sihonkpaasi*. This becomes especially evident if one asks a person who has attended church to translate the word *siyawesi* into French. The common response is *les démons* ('demons'). Others, however, translate *siyawesi* as *les nains* ('dwarfs') or *les esprits/génies de brousse* ('bush spirits').

Conclusion

> There was a diviner, a marabout (Muslim scholar) and an imam. *Uwienu* wanted to test their divination abilities, so *Uwienu* constructed a mud hut and put a cow inside. *Uwienu* then asked the *marabout* to tell him what was in the room and how it looked. The *marabout* said that there was an animal inside but he neither knew what it was nor its colour. *Uwienu* then asked the imam, who also said that there was an animal inside but he couldn't say more than that. *Uwienu* then asked the diviner who, with the help of the *siyawesi*, confirmed that there was an animal inside, that it was a cow and that it had various markings of this and that colour. *Uwienu* was impressed with the diviner and affirmed him as better than the marabout and imam. His success was due to the *siyawesi*.
>
> (Recounted by Basadi, June 2011)

This chapter has presented a case study of the *siyawesi*, the name employed for bush dwarfs by the Bebelibe of northwestern Benin, demonstrating that the *siyawesi* are important to Bebelibe ontology due to their role in divination as *Uwienu*'s (God's) messengers. Local church teaching, however, influences how people understand the *siyawesi*, with the result that some people now associate them with *Disɛnpode*, who in turn has taken on the role of the devil.

When Huber employed the word *mythique* ('mythical') (1973: 378) to characterise the *siyawesi*, it is not evident what sense he accorded the word. It could be interpreted in a number of ways: something idealised, fictitious, derived from a myth, imaginary or legendary. Regardless of what he meant by this, it is clear from what the interviewees shared with me that all these interpretations are problematic. Everybody I spoke with affirmed that the *siyawesi* exist and that they seek to have reciprocal relationships with certain people. Those chosen by the *siyawesi* often become diviners and are instructed by the *siyawesi*. Thus, the *siyawesi*'s intermediary role is not mythical but very real.

I agree with J. Merz (2004; see also Merz 2014: 216–225) and other contemporary scholars who write about witchcraft and sorcery (e.g. Nyamnjoh 2001; Stoller and Olkes 1987: 229; see also Bernard, this volume), in addition to the large body of literature emerging from the ontological turn (e.g. Henare et al. 2007; Scott 2013; Kohn 2015), that together with representing the views of a given people group, we should also 'accept the African discourse [...] as valid and affirm that, yes, witchcraft *does* exist' (Merz 2004: 576, italics in the original). In the same way, I propose that bush dwarfs *do* exist even though they elude the 'scientific verification' typically required if cryptids are to be taken seriously by external commentators.

Notes

1 The results detailed in this chapter are based on personal observation and in-depth interviews conducted either by Johannes Merz or myself. Interviews were semi-structured using open-ended questions. All interviews were conducted in French and Mbelime and were recorded. The interviews were fully transcribed and the Mbelime was translated into French. All English translations are mine. My thanks to Bienvenue Sambiéni and Claire N'Tade for their help with transcribing and back translating the Mbelime texts. I would also like to thank Johannes Merz for his support as I wrote this chapter and Samantha Hurn for encouraging me to pursue my research into the *siyawesi*.
2 All names are pseudonyms.
3 The Bebelibe and their language, Mbelime (which is Gur), are also known by the names Niendé (e.g. Maurice 1986: 4) or Nyende (e.g. Huber 1973).
4 If a person wants to speak with an animal, or vice versa, either the person or the animal needs to shape-shift.
5 The word *sanpola* means 'the cursed ones' in Moba (Koabike, personal communication, 16 July 2011).
6 *Disɛnpode*'s character is very close to that of the Moba's *sanpola*. Moba and Mbelime are both Gur languages. *Sanpola* and *Disɛnpode* have the same root: *sanpo*– and – *sɛnpo*–.
7 Children now draw pictures of animals at school. Consequently, there is currently some artwork with animals, such as a mural at the cultural centre in the town of Cobly. Photographs, especially of people, have also become important (see also Merz 2014: 202–207).
8 The same procedure is carried out for both men and women, although women rarely practise formal divination following their initial installation, as mentioned above.

References

Akibo, P.K. 1998. *The Fruits of Pentecost: The History of the Evangelical Assemblies of God Church in Benin (E.A.G.C.-Benin)*. Abomey-Calavi, Benin: E.A.G.C.-Benin.

Bannerman-Richter, G. 1987. *Mmoetia: The Mysterious Little People*. Elk Grove, CA: Gabari Publishing.

Bayart, J.F. 2008 [1998]. Fait Missionnaire and Politics of the Belly: A Foucaultian Reading. In *Readings in Modernity in Africa*, edited by P. Geschiere, B. Meyer and P. Pels. Bloomington and Indianapolis: Indiana University Press; Oxford: James Curry; Pretoria: Unisa Press, 92–98.

Bird-David, N. 2006. Animistic Epistemology: Why Do Some Hunter-Gatherers Not Depict Animals? *Ethnos*, 71(1), 33–50.

Bonnet, D. 1981–82. La procréation, la femme et le génie (les Mossi de Haute-Volta). *Cahier ORSTOM*, 18(4), 423–431.

Bonnet, D. 1986. *Représentations culturelles du paludisme chez les Moose du Burkina*. Ouagadougou, Burkina Faso: ORSTOM.

Bonnet, D. 1994. L'éternel retour ou le destin singulier de l'enfant. *L'Homme*, 34(131), 93–110.

Bonnet, D. 1995. Identité et appartenance: interrogations et réponses moose à propos du cas singulier de l'épileptique. *Cahiers des sciences humaines*, 31(2), 501–522.

Cardinall, A.W. 1927. Note on Dreams Among the Dagomba and Moshi (Northern Territories, Gold Coast). *Man*, 27, 87–88.

Cardinall, A.W. 1931. *Tales Told in Togoland*. London: Oxford University Press.

Cohen, E. 2008. What Is Spirit Possession? Defining, Comparing, and Explaining Two Possession Forms. *Ethnos: Journal of Anthropology*, 73(1), 101–126.

Comaroff, J. and Comaroff, J. 1992. *Ethnography and the Historical Imagination*. Boulder CO and Oxford: Westview Press.

Cornevin, R. 1981. *La République Populaire du Bénin: Des origines Dahoméennes à nos jours*. Paris: G.-P. Maisonneuve et Larose.

Erny, P. 1988. *Les premiers pas dans la vie de l'enfant*. Paris: L'Harmattan.

Erny, P. 2001. Préface. In *Vie, mort et ancestralité chez les Moba du Nord Togo*, by D.B. Guigbile. Paris: L'Harmattan, 11–20.

Fainzang, S. 1986. *L'Intérieur des choses: Maladie, divination et reproduction sociale chez les Bisa du Burkina*. Paris: L'Harmattan.

Geschiere, P. 1997. *The Modernity of Witchcraft: Politics and the Occult in Postcolonial Africa*. Charlottesville and London: University Press of Virginia.

Geschiere, P., Meyer, B. and Pels, P. 2008. Introduction. In *Readings in Modernity in Africa*, edited by P. Geschiere, B. Meyer and P. Pels. Bloomington and Indianapolis: Indiana University Press; Oxford: James Curry; Pretoria: Unisa Press, 1–7.

Gilbert, M. 1998. Concert Parties: Paintings and Performance. *Journal of Religion in Africa*, 28(1), 62–92.

Goody, J. 1962. *Death, Property and the Ancestors: Study of the Mortuary Customs of the LoDagaa of West Africa*. Stanford, CA: Stanford University Press.

Goody, J. 1997. *Representations and Contradictions: Ambivalence Towards Images, Theatre, Fiction, Relics and Sexuality*. Oxford: Blackwell.

Guigbile, D.B. 2001. *Vie, mort et ancestralité chez les Moba du Nord Togo*. Paris: L'Harmattan.

Henare, A., Holbraad, M. and Wastell, S. 2007. Introduction: Thinking Through Things. In *Thinking Through Things: Theorising Artefacts Ethnographically*, edited by A. Henare, M. Holbraad and S. Wastell. London: Routledge, 1–31.

Heraud, M. 2005. *Malédiction et handicap: A qui la faute?* Paris: Handicap International France.

Hill, H. 1996. Witchcraft and the Gospel: Insights from Africa. *Missiology: An International Review*, 24(3), 323–344.

Hornborg, A. 2006. Animism, Fetishism, and Objectivism as Strategies for Knowing (or not Knowing) the World. *Ethnos*, 71(1), 21–32.

Horton, R. 1971. African Conversion. *Africa*, 41(2), 85–108.

Huber, H. 1973. L'existence humaine en face du monde sacré: Rites domestiques chez les Nyende du Dahomey. *Anthropos*, 68(3/4), 377–441.

Huber, H. 1979. *Tod und Auferstehung: Organisation, Rituelle symbolik und Lehrprogramm einer Westafrikanischen initianions Feier*. Freiburg, Germany: Universitätsverlag Freiburg Schweiz.

Ingold, T. 2000. *The Perception of the Environment: Essays in Livelihood, Dwelling and Skill*. London and New York: Routledge.

Jackson, M. 1977a. Sacrifice and Social Structure among the Kuranko. Parts I and II. *Africa*, 47(1), 41–49.

Jackson, M. 1977b. Sacrifice and Social Structure among the Kuranko. Part III. *Africa*, 47(2), 123–139.

Kirby, J.P. 1986. *God, Shrines, and Problem-solving among the Anuf of Northern Ghana*. Berlin: Dietrich Reimer Verlag.

Koabike, B.J. 2003. *Religion traditionnelle chez les Moba*. Lomé, Togo: SIL Togo.

Kohn, E. 2007. How Dogs Dream: Amazonian Natures and the Politics of Transspecies Engagement. *American Ethnologist*, 34(1), 3–24.

Kohn, E. 2015. Anthropology of Ontologies. *Annual Review of Anthropology*, 44, 311–327.

Manning, P. 1998. *Francophone Sub-Saharan Africa 1880–1995*. 2nd edn. Cambridge: Cambridge University Press.

Maurice, A.-M. 1986. *Atakora: Otiau, Otammari, Osuri. Peuples du Nord Bénin (1950)*. Paris: Académie des Sciences d'Outre-mer.

Merz, J. 2004. From Relativism to Imagination: Towards a Reconstructive Approach to the Study of African Witchcraft. *Anthropos*, 99(2), 572–580.

Merz, J. 2008. 'I am a Witch in the Holy Spirit': Rupture and Continuity of Witchcraft Beliefs in African Christianity. *Missiology: An International Review*, 36(2), 201–218.

Merz, J. 2014. A Religion of Film: Experiencing Christianity and Videos Beyond Semiotics in Rural Benin. Ph.D. thesis, Leiden University.

Merz, J. 2015. Mediating Transcendence: Popular Film, Visuality, and Religious Experience in West Africa. In *New Media and Religious Transformations in Africa*, edited by R.I.J. Hackett and B.F. Soares. Bloomington: Indiana University Press, 99–115.

Meyer, B. 1998. 'Make a Complete Break with the Past.' Memory and Post-Colonial Modernity in Ghanaian Pentecostalist Discourse. *Journal of Religion in Africa*, 28(3), 316–349.

Meyer, B. 1999. *Translating the Devil: Religion and Modernity Among the Ewe in Ghana*. Edinburgh: Edinburgh University Press.

Nyamnjoh, F.B. 2001. Delusions of Development and the Enrichment of Witchcraft Discourses in Cameroon. In *Magical Interpretations, Material Realities: Modernity, Witchcraft and the Occult in Postcolonial Africa*, edited by H.L. Moore and T. Sanders. London and New York: Routledge, 28–49.

Ovesen, J. 1990. Initiation: A Folk Model among the Lobi. In *Personhood and Agency: The Experience of Self and Other in African Cultures*, edited by M. Jackson and I. Karp. Uppsala, Sweden: Acta Universitatis Upsaliensis, 149–167.

Pels, P. 1998. The Magic of Africa: Reflections of a Western Commonplace. *African Studies Review*, 41(3), 193–209.

Pels, P. 2003. Introduction: Magic and Modernity. In *Magic and Modernity: Interfaces of Revelation and Concealment*, edited by B. Meyer and P. Pels. Stanford, CA: Stanford University Press, 1–38.

Piot, C. 1999. *Remotely Global: Village Modernity in West Africa*. Chicago and London: University of Chicago Press.

Rattray, R.S. 1927. *Religion and Art in Ashanti*. Oxford: Clarendon Press.

Scott, M.W. 2013. The Anthropology of Ontology (Religious Science?). *Journal of the Royal Anthropological Institute* (n.s.), 19(4), 859–872.

Stoller, P. and Olkes, C. 1987. *In Sorcery's Shadow: A Memoir of Apprenticeship among the Songhay of Niger*. Chicago and London: University of Chicago Press.

Tengan, E.B. 1990. The Sisala Universe: Its Composition and Structure (An Essay in Cosmology). *Journal of Religion in Africa*, 20(1), 2–19.

van Dijk, R. and Pels, P. 1996. Contested Authorities and the Politics of Perception: Deconstructing the Study of Religion in Africa. In *Postcolonial Identities in Africa*, edited by R. Werbner and T. Ranger. London and New Jersey: Zed Books, 245–270.

7 Suspending disbelief and experiencing the extraordinary

How radical participation may facilitate an understanding of aquatic snakes and fish-tailed beings in southern Africa

Penelope Bernard

Introduction

A typical feature of cryptozoology, or the study of hidden, secretive creatures (cryptids) not officially recognised by science, is that it is usually only considered seriously by a fringe element of society that draws its evidence from anecdotal sightings of what are assumed to be zoological entities that are physically tangible; their confirmation as a recognised species being dependent on their capture, or their potential for leaving physical remains as evidence, or by independent verification of sightings by a number of sober minded scientists (see Forth, this volume; Walsh and Goldman, this volume). Cryptids are often suspected of being vestiges of species thought to be extinct but have somehow managed to survive human categorisation and domination due to their shy or elusive natures. They are brought to the forefront of public consciousness and imagination when rare, and often brief sightings are claimed by individuals or small groups who happen to be traversing their territories. While public imagination may be stimulated by such claimed sightings (Dendle 2006), serious scientific consideration is usually frowned upon and belief in cryptids tends to be viewed as occupying the realms of psychological fantasy or disorder (Sharps et al. 2010).

The range of creatures falling into the category of cryptids is vast (Eberhart 2002; Newton 2005), and includes those species not officially recognised by science. These may range from species thought to be extinct or unknown, to creatures out of place in that they furtively occupy territories not their own, to mystical and magical shape-shifting creatures that transcend the various orders of reality as we understand them. These different categories require different sorts of analysis and understanding and thus some classification of cryptids is in order. A discussion on classification is not my intention in this paper (see Turner, this volume); my interest, however, is with the last of the categories mentioned, those phantasmagorical and shape-shifting creatures often associated with fantasy, in particular shape-shifting water serpents and their associates, fish-tailed beings or mermaids (see also Schmidt, this volume). I argue that research tools beyond those typically associated with scientific verification may be required in order to understand claimed human experiences of them – experiences which appear to transcend time and space and form the basis of many religious belief systems. To support my argument I draw from my own experiences of radical participation amongst Nguni-speaking diviner-healers (Zulu = *izangoma*/pl; Xhosa= *amagqirha*/pl) in South Africa, which has allowed me to glimpse the nature of such beings from a different perspective or 'order of reality' (Schutz 1974) to that of science, which presently does not have the tools to address such encounters.

Radical participation and the extraordinary

In 1997 I was informed during a divinatory session with a powerful Zulu diviner-healer (*isangoma*), whom I refer to as Baba, that I had a calling to be an *isangoma*. This was my first visit to Baba, whom I hoped to interview regarding the claim that he had been taken by a giant mystical snake (experienced in a trance state) under the sea for three hours off the coast near Durban in KwaZulu-Natal, South Africa. During the introductory divination he reported that I was connected to the water divinities (the snake/mermaids), whose power resided in a river on the farm on which I grew up as a child in the Eastern Highlands of Zimbabwe (Nyanga), which is located over 1,500 kilometres north of Durban. I subsequently discovered that the area around the farm abounded with reported sightings of mermaids (*njuzu*) and powerful pythons and/or shape shifting snakes. Baba insisted that if I wanted to fully understand the experiences of those who claimed to have such encounters, or of being taken underwater (like he had been), I needed to accept the calling and experience it from the inside. As I had experienced some instances of precognition in the previous year, and certain events had been happening that were challenging my sense of reality, I was ready to suspend my disbelief and engage in what has been termed radical participation and the anthropology of extraordinary experience (Young and Goulet 1994).

I was subsequently initiated as a novice into Baba's training school, and after having certain ritual inductions performed, which included being given holy water and certain medicines (roots and barks) to drink and purge with, I started to experience occasional very clear and profound dreams. Some of these dreams involved encounters with pythons, large non-aggressive snakes, and fish-tailed beings. I accept that the nature of these dreams, which did not appear to be dose related, could be explained as being induced by either auto-suggestion or the psychoactive properties of the plants administered (see Attala, this volume). These water serpent/mermaid or 'nixie' themes have been identified by psychologist Carl Jung as universal archetypal experiences that arise from the collective unconscious (Jung 1968), and the religious historian Mircea Eliade (1991) has also noted these frequently recurring water/serpent/mermaid themes, which are found in a number of different shamanic traditions. Eliade refers to these as 'aquatic hierophanies' (see also Bernard 2013) and, as with other religious symbols, he proposes that they reveal 'certain aspects of reality – the deepest aspects – which defy any other means of knowledge [...] They bring to light the most hidden modalities of being' (Eliade 1991: 12).

While these phyto-induced experiences might serve as supporting evidence for Jungian psychology, or Eliadian symbolist hierophanies, what perplexed me was that the dreams offered details of the geographical location of where these aquatic deities were to be found, none of which I had ever known existed or had been to in physical reality. My task, as instructed by my teacher, was to locate the pools shown in my dreams and to visit them so I could 'show' myself to the divinities and give offerings of thanks for the dream callings I had received.

Discovering the location of these pools, lakes and springs was frequently convoluted and depended on a series of synchronistic events which often took a number of years to unfold; my subsequent discovery of the sites revealed remarkable similarity to the geographical features I had seen in my dreams, as well as some of the people encountered in them. Most of the dreams sites I have been able to discover (five) were located across southern Africa, and three were located outside of Africa. It must be emphasised that I had never visited any of these sites prior to these dreams, and at all the sites that I have subsequently been able to locate following the dream, I have found, on discussion with local inhabitants, one or more of the same recurring themes. These include: reported sightings of large mystical snakes and/or mermaids or fish-tailed beings; rare instances of submersion or disappearance of certain individuals underwater for long periods of time after which they emerge with powerful abilities in divination; reports of occasional drowning or the death of individuals due to certain precautionary rules not being adhered to; the practice of ritual offerings to appease or appeal to the water divinities; and/or the existence of myths relating to such themes. I only saw these snake/mermaid creatures in my dreams, and not at the physical sites that I subsequently discovered, but the coincidence of finding the exact place shown in a dream (previously unknown to me) and the concurrence of reported sightings and/or related myths, have prompted me to take such reports seriously.

I have already discussed one of these dream experiences in another publication (Bernard 2008) where I recounted how I was led, through a series of dreams, to a pool in the midlands of KwaZulu-Natal reputed by local inhabitants to be the abode of the Zulu princess of heaven, *Inkosazana*, sometimes referred to as *Nomkhubulwana* (who some claim can manifest as a mermaid); in this present chapter I will draw on another of those experiences located over 1,000 kilometres from the site mentioned above, to a remote river pool on the Doring River in the northwest region of the Western Cape Province of South Africa. The fact that I had never been in this area, nor had a family history associated with it, and the fact that this territory is occupied by Khoesan[1] descendants, and way beyond the reaches of my Zulu teacher's realm, is pertinent. The accuracy of this dream, which I return to later in the chapter, concurred with many of the elements that I subsequently discovered at the site and has further challenged my rational understanding of claims of snake and mermaid sightings. In line with the emphasis of the turn to ontology in anthropology more generally (e.g. Kohn 2015; Viveiros de Castro 2015), I argue that we need to consider more broadly what constitutes evidence when dealing with phenomena that require one to move beyond the 'finite provinces of meaning' that constitute the order of reality in which scientific verification is currently located (Schutz 1974).

Transcending orders of reality and Schutz's phenomenological conundrum

The anthropology of extraordinary experience (AEE) is primarily a methodological approach which advocates that when an anthropologist is seeking to explore phenomena or claims that go beyond our 'normal', 'rational' ('Western scientific') understanding of reality, especially those relating to altered states of consciousness (which include dreams, trance states and visions), we should suspend our disbelief and allow ourselves to experience such claimed phenomena from the inside. The AEE approach, which advocated the inclusion of such extra-ordinary experiences as valid data, was first articulated in a publication edited by Young and Goulet (1994) entitled *Being Changed by Cross-cultural Encounters: The Anthropology of Extraordinary Experience*, and this was followed by a sequel edited by Goulet and Miller (2007) entitled *Extraordinary Anthropology: Transformations in the Field*. Drawing from their own experiences they note that when ethnographers have extraordinary experiences in the field, these 'challenge one's conception of reality in the sense that normal ways of classifying perceptual data are no longer adequate and the boundary between the real and imaginary are blurred' (Young and Goulet 1994: 8). Emphasising the value in gaining an insider perspective of extraordinary experiences, Young and Goulet argue that 'anthropologists should, at a minimum, temporarily suspend disbelief, and attempt to take as seriously as possible informants' reports of extraordinary experiences, as well as their explanations for them' (1994: 11). The proponents advocated that in order to gain better insights into various realms of spiritual or metaphysical knowledge, the ethnographer should become actively involved, through 'radical participation', or more specifically 'radical embodied participation', in the ritual life of his or her host society. Rituals can be the best means of 'radical embodied participation' through which altered states of consciousness, including powerful dreams and visions, are induced.

The concept of 'radical participation' that was used by Goulet and Miller (2007), was adapted from Fabian's concept of 'ecstatic anthropology' (2001: 31), which advocates that anthropologists set aside their 'single minded pursuit of data within a clearly defined research agenda' (Goulet and Miller 2007: 1) and allow themselves to become *co-actors* in their informants' worlds. It thus goes beyond our mere observing, questioning and listening, to actively participating and 'joining in' with their lives. The proponents of this approach also insist that, as anthropologists, we cannot set aside those experiences 'that challenge our own epistemological, ontological, and ethical assumptions' (Goulet and Miller 2007: 2). They clarify this statement by referring to Fabian's concept of ecstatic fieldwork, which offers the potential for anthropologists to 'step outside one's taken-for-granted body of knowledge' (Goulet and Miller 2007: 5). Allowing one's world view to become transformed in the process is a distinct aspect of such an approach and to 'include such moments (of transformation) in one's ethnographic account is to present not only knowledge obtained but also the processes through which such knowledge was gained and the circumstances in which such processes became operative' (Goulet and Miller 2007: 6). Michael Jackson (1989) has used the concept of 'radical empiricism' to argue much the same point. Jackson drew his insights from William James (1976), a philosopher who coined the term 'radical empiricism'. Radical empiricism was, first and foremost, 'a philosophy of the experience of objects and actions in which the subject itself is a participant' (Jackson 1989: 3, citing from Edie 1965: 119). The selection of the term 'radical empiricism' was employed to directly counter the existing concept of 'traditional empiricism'. The latter 'assumes that the knower and the known inhabit disconnected worlds and regards experience as something passively received rather than actively made, something that impresses itself upon our blank minds or overcomes us like sleep' (Jackson 1989: 5). Thus, radical empiricism breaks down the boundary between the observer and the observed. Radical participation, as a methodological concept (Goulet and Miller 2007) could be seen as an element of this notion of radical empiricism.

I argue that the present reification of scientific validation of phenomena or claimed sightings of cryptids, in order for them to be considered 'real', is problematic. Phenomenologist Alfred Schutz argued that experiences that take place in different 'orders of reality' such as in altered states of consciousness, as in dreams and in trance, are governed by 'finite provinces of meaning' that are applicable only to the order of reality in which they are experienced and understood. According to Schutz, these orders of reality (such as the everyday life-world, the world of dreams, the world of science or the world of religious experience) each have their own meaning-compatible experiences, or cognitive styles (Schutz 1974). Schutz argued that one cannot merge the cognitive styles of the different provinces of meaning but that one should rather 'leap' across them. However, there is a residue (what Schutz referred to as an 'enclave') left behind from an experience gained in one order of reality (such as a dream) that bears relation to another (our everyday life-world). In the case of a dream experience we can only measure its truth or falsity within the cognitive style of a particular order of reality from which we are examining it (Schutz 1974: 127); as scientific knowledge does not derive from the order of reality in which dreams or trance states emerge, it cannot be used as a valid measure of their experience. If Schutz's position is correct, this makes any 'scientific verification' of the claimed reality of dream-based experiences impossible, unless science is able to develop conceptual tools that are compatible with the order of reality that governs dreams. While not precluding the value of including such experiences as valid data in ethnographic accounts, or subjecting them to psychoanalysis, this phenomenological principle poses a problem if one is trying to make 'scientifically valid' claims for experiences or beliefs that are derived from dreams or during other altered states of consciousness.

While these arguments and the conundrums they pose are pertinent to experiences relating to the dream world or trance states (as 'altered states of consciousness' or ASCs), the phenomena I deal with in this paper include claims of ASC experiences, but they also transcend them to include claims of physically tangible encounters with snake and/or mermaid beings in geographical habitats that challenge our concepts of time and space. There is often a meaningful enclave or residue that binds these different experiences together, and this convergence of spatio-temporal dimensions across different orders of reality is an important aspect of extraordinary experience. The order and timing in which events occur (be they of dreams, encounters with people and/or animals, rituals, messages, etc.), and the convergence of otherwise disparate events, can transform an ordinary 'random' event into one of great significance and can have an effect of heightening the intensity of the experience. Goulet and Miller (2007) observed this spatio-temporal aspect of extraordinary experience in their review of contributors' papers in their edited volume. This tendency towards 'a sort of non-linear, non-Newtonian, quantum physical world in which related, connected events unfolded [or appeared to unfold] simultaneously in various locations' (Goulet and Miller 2007: 4) is sometimes referred to as 'synchronicity' (Jung et al. 1964; Mansfield 1995). Mansfield (1995), a physicist, has sought to explain the phenomenon of synchronicity with a combination of Jungian psychology and the insights of quantum physics. He defines synchronistic experiences as 'numinous events where the outside world meaningfully relates to our inner psychological states' (1995: 6). Jung's understanding of synchronicity differs slightly however in that he defines it as 'a "meaningful coincidence" of outer and inner events that are *not themselves causally connected*' (Jung et al. 1964: 211, emphasis added). For believers in the metaphysical power of these beings, as found in southern Africa, this definition of Jung's could be problematic in that they would see the coincidence of dreams, events, and encounters across different orders of reality, as well as the disturbance of our normal understandings of the boundaries of time and space, as having intimate causal relations. In the next section I examine how encounters with snake and fish-tailed beings are understood by diviner-healers in southern Africa, and these include the different orders of reality in which they are believed to manifest and transcend.

Water snake and mermaid encounters in southern Africa

Knowledge and beliefs regarding shape-shifting water serpents and mermaids are deep-seated and extensive among many southern African diviner-healers who frequently report encounters with such creatures in both their everyday life-worlds or in altered states of consciousness, such as in dreams or trance states (see Bernard 2010 for extensive comparative research of these beliefs across southern Africa). Far from being anecdotal, these claims are buttressed by a complex system of ideas and ritual practices regarding them. Reports of such beliefs have been documented from the time serious ethnographical enquiry was launched in southern Africa in the early to mid-1800s, largely by missionaries and ethnographers such as Callaway (1970), Soga (1931), Bryant (1949), and more recently Berglund (1976). The longevity and complexity of these ideas and the knowledge and rules regarding the ways humans should act towards these water creatures and the natural resources they inhabit and control, makes the common argument that these ideas originate from European representations of these entities (Drewal 2008) unlikely.

In southern Africa these water snake and mermaid cryptids are seen as working alongside human ancestral spirits (which are also believed to manifest as snakes, birds or other animals) to provide certain chosen individuals with insight into past, present and future events and in providing skills of healing (the role of diviner-healers). Explanations of the role and function of these fish-tailed and snake beings bear all the hallmarks of divinity, in that they are seen as the forebears of humanity (if not life), the source of water, rain, fertility, and the guardians of morality. Claimed experiences of them range from significant actual encounters with normal sized or giant light-emitting snakes and/or fish-tailed beings at water sources or remote natural sites (mountains, forests), to dreams, or to the ultimate of all experiences, the physical submersion of a individual 'called' underwater for varying periods of time (from a few hours to months or even years). The successful return of the submerged individual is dependent on the correct actions and behaviour of their kin members, who must not cry or grieve at the disappearance or apparent drowning of their family member but must wait patiently for signs of the submerged individual's return. They are usually required to offer a goat or cow as a sacrificial exchange for the life of the victim, who will then re-emerge, bestowed with powerful divinatory abilities and knowledge of healing and/or medicines. Individuals who undergo such submersion and return to tell the tale often report that they were confronted with a giant snake, a fish-tailed woman, multiple other snakes or creatures, and their own deceased kin (ancestral spirits) who occupy an idyllic dry world beyond the water realm.

These divinities are believed to traverse the boundaries of reality from the everyday life world and its physical elements (especially water) to the world of spirit (be it the underworld or heavenly realms). They are thus viewed as being physically tangible but have the magical ability to make themselves invisible and cross the boundary into spirit. They allegedly have shape-shifting and transformative abilities, with the snake being described as able to emit an electrical force-field, often in the form of a bright light emanating from its eyes or a crystal stone on its forehead, which can trap the gaze of a person and put them into a trance before taking them to the underwater world.

The snake is believed to be able to change its form, as well as its size and length, sometimes concealing its head or tail, and some claim it emits a rustling sound when encountered. Some informants claim that the snake transforms into a mermaid and vice versa, to the extent that they are sometimes seen as one and the same, while others claim that the snake/mermaid duo are the gendered aspects of God (an androgynous God). For some Zulu informants who claim to have encountered the snake in its underwater realm, they explain that this is how the sky God *Inkosi yamakhosi omkhulu* ('the lord of lords') manifests, and the mermaid is none other than his daughter, the heavenly princess/goddess *Inkosazana* (or *Nomkhubulwana*), who some see as the source of life, water and fertility (Berglund 1976; Bryant 1949; Kendall 1999). It is asserted that *Inkosazana* can transform into a tree, a plant, a small animal, a snake, a beautiful young woman, an old hag, light rain/mist or a rainbow.

Rather than viewing the mermaids as a singular goddess, the Xhosa and Shona speakers regard them as a group of powerful semi-daemons (referred to as the *abantu bomlambo*, or 'people of the river' by the Xhosa, and *njuzu* by the Shona) whose role in the generation of water, fertility and healing is much akin to the Zulu ideas of *Inkosazana/Nomkhubulwana*. Ideas of the water snake and the *watermeid* ('mermaid') are also extensive across the southern Cape region amongst the original inhabitants, the Khoesan peoples and their descendants. Numerous claims of sightings and encounters with the snake and the *watermeid* emanate from these areas (Hoff 1997; 2007), and are frequently reported in the popular press. I have argued (Bernard 2010) that beliefs and practices regarding these water divinities held by the Nguni-speaking peoples (Xhosa, Zulu and Swazi) could well have been shared with, or possibly originated from, the Khoesan (San and Khoekhoe) peoples with whom they intermarried and co-existed along the eastern seaboard of South Africa.

Reported sightings of the mermaid, usually as a female entity, reveal remarkably similar descriptions across southern Africa (from Zimbabwe and across the whole of South Africa) in that she is described as having fair skin (European-like – see also Schmidt, this volume), long (wet) dark hair and, some report, green/blue eyes. Dreams are one of the main ways in which these divinities are believed to communicate with the living, often calling individuals to certain pools or water bodies where they are reported to live. For diviner-healers who receive such dreams it is important for them to try and locate these pools (clues to their location being given in the dream), where they must visit and 'show' themselves by giving offerings of sorghum beer, beads, tobacco and on occasion small stock (such as chickens or goats). These mermaid/snake divinities are associated with certain undisturbed pools and fountains, but it is believed they can move across the landscape in the form of winds (including tornadoes) mist and rain.

While both the snake and mermaid are essentially positive in their creative and moral force, they can be exceedingly destructive when offended or angered and are believed to seek revenge for bad human behaviours (for social and environmental reasons) by inflicting them with extreme weather events. Among the Khoesan descendants it is also believed that certain individuals are able to trick the snake, or placate it with herbal medicines, in order that they may remove the powerful crystal stone from its forehead (Hoff 1997; 2007) to use for healing or luck. Among the Nguni-speakers there is also widespread belief that certain magicians (herbalists) are able to mystically trap these beings through obtaining part of their flesh and using contagious magic to compel them to enter into an individual exchange pact whereby they, or their clients, can gain rapid and enormous wealth (Bernard 2010), usually in exchange for the life of a close kin member. Such ideas, now rife in the context of the disruptions caused by neo-liberal economics, support the claim that these beings can be physically tangible (cf. High, this volume). However, discourses regarding the magical manipulation of these creatures reveal that such activities are not only morally suspect but also highly risky, as these divinities have consciousness and agency that exceeds that of humans and they can punish evil doers with death. The physical tangibility of the fish-tailed beings is also evident in the claims by certain communities that occasionally misguided individuals have been responsible for shooting and killing a mermaid, actions which have resulted in devastating floods with multiple casualties (Bernard 2010; Pienaar 2009).

Apart from these rare occasions it is generally believed that if the water divinities choose not to be seen they will not reveal themselves and this is especially the case if a person's motives are insincere or morally suspect (Bernard 2008). It is believed that such people run the risk of being killed if they trespass too near to the divinities' watery abodes. It is for this reason that most encounters happen to young children (who are seen to be less tainted by bad deeds or intentions) or diviner-healers. It is also believed among the Zulu that certain kings or chiefs may commune with them.

The seriousness with which local indigenous populations on the southeastern seaboard of South Africa take alleged sightings of these water divinities was demonstrated in two alleged hoaxes in the 1990s. In 1996 in Oudsthoorn, in the Little Karoo, South Africa, the CP Nel Museum reported it had captured a mermaid (referred to as the *watermeid* by local coloured or Khoe descendants) which it was holding in a water tank. This alleged discovery of a beached mermaid coincided with a time of severe flooding in the region and led to a deluge of enquiries and visitors, with some people travelling hundreds of kilometres to see this captured creature with their own eyes. When the museum claimed it had released the mermaid into a nearby river and had installed a mannequin mermaid in the tank instead, there was great dissatisfaction. A similar response was elicited when the *Daily Dispatch*, a local newspaper serving the East London area of the Eastern Cape, published a report on 1 January 1998 claiming that a stranded mermaid had been rescued from the beach and relocated to the city's oceanarium where it could be viewed by the public (for a fee); when it became apparent that the 'mermaid' was a publicity stunt the angry crowds who had travelled long distances became very upset, demanding refunds for their travel costs (Morrow and Vokwana 2003).

While both claims turned out to be hoaxes, it was the flood of interest and strong emotional reactions that they generated that is of interest. Indeed, the popular press frequently carries such stories of mermaid sightings in South Africa, and many of these seem to come from the Khoesan associated territories of the Little Karoo. An example of such a story comes from an article published in the *Sunday Times* (24 November 2002) by journalists Caspar Greeff and Ruben Boschoff, who recounted their findings when they went to the Baviaanskloof, a wilderness area in the Eastern Cape. They tell of their meeting with an old medicine man, Oom Klaas, who had reported sightings of mermaids. He told them that many people living in the area had seen them, and claimed:

'I saw the mermaids twice. The first time was years ago, in a pool near here. She had pitch black – long hair – and pure white skin and breasts like a woman. The bottom half of her body was in the water. She was combing her hair with a black comb, and when she saw me, she went underwater and disappeared. As she went under her hair spread out on top of the water. It was beautiful. I saw another mermaid two years ago in the same pool. Her breasts were bigger, she looked like she had had a baby, but she was very beautiful. She looked at me – she had grey-green eyes – then she also disappeared under the water.' Oom Klaas had been warned about the mermaids by his father, also a herbalist and medicine-man […] 'My father saw many mermaids. One Sunday he went to the pool and sat on the rocks and they took him down to where they live under the water, and taught him about herbs. They live in houses like we do, only under water […] Sometimes they will pull a child under the water. If that happens to your child, you mustn't cry and carry on, because then they will kill the child and throw her out (of the water). No, you must get a cow, and slaughter it next to the river, and cook it. Then send down the haunches. When you go home your child will come back.'

(Greef and Boshoff 2002)

This report, which bears all the typical themes of the water divinities found elsewhere in the country, came from a remote area on the east coast of South Africa's southernmost regions. I was to discover similar ideas near the west coast, some 600 kilometres away.

Challenging boundaries – mermaid dreams and an anthropologist's experience

The forty degree Celsius February heat was pulsating through my body and across the landscape to the shimmering blue water of the Doring River, with its pools encased with tall reeds and mirroring the looming sandstone cliffs along its edge. I was standing, overlooking the river valley, at the very farm-workers' cottages that I had seen in my dream, exactly as they had been aligned in a row along the base of a small mountain. I was having trouble distinguishing whether I was still in my dream state or in my normal ordinary reality. The dryness of the landscape and sparse Karoo vegetation was exactly as I had sensed in my dream, as was the river. I was standing listening to Joy, my Afrikaans-speaking colleague who had kindly agreed to travel this long distance with me to interview some of the local people in their language regarding the possible significance of me being shown this place in a dream. Joy was deep in conversation with a middle-aged man, Jacob (pseudonym), and his wife about some of the local stories regarding strange encounters with cryptozoological water beings in the area. I had discovered this dream site a month earlier, but before I continue, more explanation is needed regarding my dream and how I came to find this site.

The dream, which I had in February 2004, was as follows:

> In the dream I was seated in the front of a truck with my sister and brother and we were driving along a dirt road. I was very clearly shown the type of vegetation, which was typical Karoo veld, with its short tufts of wheat coloured grass and small scrubby shrubs and the general feeling of dryness of the soil. Suddenly we came to the edge of the plateau we were driving along and the beautiful vistas of distant hills and valleys opened out in the view before us. It felt like we were at the top of an escarpment or pass, and I remember coming to a screeching halt in the car because it felt like we were about to go over the edge, and I was relieved that I was able to stop in time. As we sat in awe of the beautiful view before us, I then saw that the road veered slightly to the right, as it dropped down into the valley. I felt it was safe to continue our journey and we started to descend slowly down the steep rocky track. On our left was a sheer drop and I drove carefully, hugging the right hand side of the road. In the distance I could see a large river meandering, but it did not seem to have much water in it. We then found ourselves in the valley and we had stopped at a small isolated farmhouse. I then found I was inside the house and was aware of the presence of its owners, who were a middle-aged white couple. I noticed in particular the man, as he sat in a chair in the corner of a dark room, and looked at me with a serious face. The next moment I felt I was outside again and had stepped straight into some deep water in a river. I was able to grab hold of a rope and swing myself across to the water's edge. I noticed the old man just sitting watching me, fairly unconcerned. I threw the rope to my sister and saw that my brother (who in real life is a keen ornithologist) was already on the other side of the river watching birds through his binoculars amongst thorn trees. I then found myself in the courtyard of the house and saw a woman standing next to a sculpture of a mermaid. I was aware she was

using it as an attraction for tourists to come and visit the area. Suddenly a woman with a mermaid tail went past me, her tail flapping the ground. As I looked up I saw a small distinctly shaped hill, not far from the river, rising up out of the valley against which was a group of about six to ten modestly sized houses. I asked the lady where I was and she told me 'You are on the road to XX'.[2] I awoke with the name reverberating in my mind.

Awaking from the dream, I immediately went to consult my South African atlas. I looked for the name of the place the lady told me in the index (I was not familiar with the name), and found a place of that name with the map reference. Turning to the correct page I discovered the place was on the Doring River, in the Western Cape, in a region north of the Cederberg Mountains. I noticed there were a number of passes and escarpments marked in the area and thought that I needed to follow-up on the dream and see if the place was anything like what I was shown. This was a region of South Africa I had never been to, but as I had heard about the beauty of the Cederberg Mountains, I decided to persuade my family to accompany me in search of my dream site during our Christmas vacation that year. I made a point of telling them my dream in great detail so they could be my witnesses should we find the place that fitted my description, and in early January the following year we took ourselves off for a camping trip to the Cederberg.

Accompanying me were my brother and sister who were with me in the dream, my sister's two sons, my two younger children and a couple of their friends. It was exactly as I described, in terms of the scenery, the pass, the river valley, the single farmhouse and the special shaped hill along which was grouped a number of small labourers' cottages. Unfortunately, not one of us in our party could speak Afrikaans, the local lingua franca, to find out if there were any stories relating to mermaids in the area. All our efforts to communicate with the local 'coloured'[3] people living in the area came to nought. I was reluctant to go and knock on the farmer's door, as I feared, quite rightly, he would think I was mad if I told him that I had been shown his house in a dream. I was happy enough to know that I had found what I regarded as the right place and my brother kindly accompanied me back to the river site at dawn the next day in order that I could do a small ritual. Before descending to the river we had stopped at the worker's cottages to ask a lady if it was alright to walk down to the river. Through gesticulations and smiles the lady, who lived at the end of the row of worker's cottages (that had appeared to me in my dream), indicated the best route down to the river where I was able to access a large pool surrounded with reeds.

It was this lady and her husband Jacob that Joy and I were now interviewing to find out if the area had any connection to water snakes and/or mermaids. The lady had remembered me, and when Joy had told them I had been shown this place in a dream, far from being surprised, they beamed with smiles and appeared to accept it as if such an event was normal. Jacob then began to recount his own experience with what he referred to as the *waterbaas* ('boss/lord of the water'), which bore many of the recurring themes I had encountered across southern Africa. He explained to Joy that as children they had always been warned by the elders to avoid going close to the river because it was inhabited by a large water snake with a diamond on its forehead that would suck them down to the watery depths. He remembered that they often saw a white light near the river at night, often accompanied by the sound of cattle lowing, which was confirmation that the snake was active. He was told a person could steal the light if they were clever, but if the snake knew of your intentions then a sudden wind would come up and one had to be very strong to withstand it.

He then explained a more recent incident he had experienced as an adult, which made him take such warnings seriously. One of his responsibilities on the farm was to manage the water pump which was located on the bank of the river, near to the bridge. One day he had gone down to check the pump and as he approached it he saw a very large 'fish' lying on the sand bank. It was so big he thought he should hurry back to his car to get his gun so he could shoot it. As he returned from the car, gun in hand, the fish dived into the water (rather than slid) and at the same time an extremely strong wind (like a whirlwind) suddenly picked up on the side of the bridge where he was standing. The thick reed beds on the river's edge were flattened by the wind, but he noticed that on the other side of the bridge there was no wind at all. When he got to the pump he discovered that the wind had wrecked it and all the bearings and seals were broken. He described how this *waterbaas* was not a typical fish with a pointed face but had a flattish face. He apologised to us that this was his only experience with a water creature, but that we should seek out an old shepherd who lived upstream of them who had many stories regarding local lore. He suggested we should first stop off at the white farmer's house to ask for directions.

Encouraged by Jacob assuring us that the white farmer was a kind man and would be quite tolerant of such enquiries, we then drove to the house and knocked on the door. A middle-aged man, who bore striking resemblance to the man I had seen in my dream, answered the door and on hearing about our research interests invited us in (we did not divulge my dream to him nor mention it). As we sat in his small darkened lounge I had the same sense of déjà vu sweep over me; it was exactly like being in my dream. As he sat across from us he agreed that stories of mermaids and large mystical water snake sightings were indeed common to the area and he had heard them since he was a child. Then, in what could be construed as a perfectly rational scientific argument, he started to explain to us how a creature such as a mermaid (which he also referred to as *waterbaas*) could exist. In his opinion it was quite possible that when a mature human male was swimming in the river, he could ejaculate sperm in the water and that these sperm could fuse with any fish eggs that may be in the vicinity, resulting in offspring that were half-human and half-fish. It was evident that he had spent some time considering how one could credibly account, in 'scientific' terms, for the existence of such a perplexing creature.

After some time discussing the subject he agreed that we should interview Oupa ('grandfather'), the old shepherd (a Khoe descendent) who lived on the farm upstream. He consented to drive us to meet Oupa and introduce us to him. Initially Oupa was reluctant to divulge any information, despite his wife's obvious excitement at our presence and our topic of interest. It was when Joy mentioned to him that I had the dream that led me to this place that a smile beamed across his face and he started to open up. Like Jacob, he began by telling us that, as a child, he had always been warned not to go near the river because a large snake lived there. He was warned that if the snake caught him he would be taken under water to its 'room' (implying a dry space where it lived). The snake would normally only be seen at night (living in its room underwater during the day). His wife interjected saying that sometimes during the day, when their normally drought stricken land was about to receive good rains, the light of the snake would be seen on the mountain, much like the sun shines off a mirror. The appearance of the water snake in this manner was a forewarning that rain was coming.

Oupa then started to recount his own experience with the snake, which he also referred to as the *waterbaas*. He pointed downstream towards the pool in the distance, the very pool that I had visited a month earlier with my brother to do the ritual offering, and explained how one night he had been walking home along the edge of the river when he had observed light moving around as if someone was walking with a bright torch. However, the light was coming from the semi-submerged reeds in the water. He suspected that it might have been the water snake feeding, and as he came closer the light disappeared. Taking as wide a berth as possible he returned home, and the next morning he went down to the pool to investigate. He described how he found a *pol*, describing this as a circle of reeds with water in the middle (suggesting he had not seen it previously). When he looked inside the *pol*, he saw a deep hole in the midst of which stood a *willerhout* ('willow tree'). For some reason this tree made him very fearful and he beat a hasty retreat. As he was recounting this story his young daughter came outside to join us, and on hearing about our interests, agreed that she knew a number of people who had reported seeing a *watermeid* ('mermaid') when crossing the bridge over the river.

Conclusion

In this chapter I have drawn on Schutz's phenomenological conundrum to explain the limitations of science in its current format in validating certain types of experience that straddle different orders of reality. I have focused in particular on the widespread claims across southern Africa of encounters with mystical shape-shifting water snake and mermaid divinities. Seeking explanations for these alleged experiences within particular orders of reality, such as in dreams, or in encounters in the everyday life-world, or during the underwater submersion trance state, could be rationalised through a number of different social and psychological explanations, such as those offered by Jung.

However, it is the synchronistic connectivity between these different states and the convergence of experiences across a wide spatio-temporal dimension that poses the greatest challenge to these more positivist and reductionist explanations. Beyond Schutz's phenomenological conundrum, a major problem with validating a dream, as with any other form of ASC, is that it is an intensely private experience that cannot be shared in the same form by others. It also suffers the risk of only being partially recalled by the individual dreamer, and may be subjected to some distortion on recall due to memory loss or auto-suggestion. Strategies used by diviner-healers to minimise such distortion or memory loss include enhancement of dreams through ritual plant ingestion (dream incubation), dancing and sleep disruption, and even undergoing physical deprivation. Adopting the practice of dream sharing, preferably with a number of unbiased and independent observers, assists in validating the experience, especially if the dream is recounted in detail and it turns out to be accurately predictive in those details. The verification of those details by independent observers located at the dream site further helps to validate the accuracy of the dream, as was the case with the example I have presented in this chapter.

I have drawn on my own experiences derived from a radical participation approach to demonstrate its potential value in providing a form of experiential verification that would not normally be considered as valid in conventional science. The convergence of my dream and its many component elements with what I encountered at the Doring River is just one example of a number that I have had in the last fifteen years. The fact that the essential elements of the snake/mermaid 'mythology' coincided with those that I have discovered across a diversity of groups in the southern African region adds further credence to these claims. The profound experiences I have had have challenged my own scientifically informed understanding of reality, temporality and causality, to the point that they have deepened the mystery of these cryptozoological beings. I believe that this is where their power lies and why they have such a profound and pervasive effect on the beliefs and practices of many southern African peoples.

In anthropology we have realised that there are certain forms of knowledge and claims of experience that are not amenable to scientific testing and verification in that they confound the basic principles on which science is currently based. Trying to prove the truth or falsity of such claims may be a futile exercise. Beyond needing to interrogate why we insist that science is the only acceptable paradigm to validate sightings of cryptids in order for them to be considered real, we also need to consider whether such mysteries should be solved, or whether they should remain elusive and beyond the control of humans in this world (see also King, this volume).

Acknowledgements

My sincere appreciation goes to all those who have made this research possible and for their support in my fairly unconventional form of research; this includes my family, colleagues and the *izangoma* who have allowed me to glimpse into their world. Special thanks go to my colleague Dr Joy Owen who accompanied and assisted me at the Doring River. My deep gratitude goes to Samantha Hurn for asking me to contribute to this important, groundbreaking book and for all her patience, support and encouragement in the process. My grateful thanks go to all those who have funded my research over the years. These include the Rhodes University Joint Research Committee, the Ernest Oppenheimer Trust, and the National Research Foundation of South Africa. The ideas and opinions expressed in this chapter are not to be attributed to them.

Notes

1   The Khoesan (sometimes spelt Khoisan) are comprised of the San (or Bushmen) hunter-gatherers, the recognised autochthons of the southern African region, and the closely related Khoekhoe (also spelt Khoikhoi), who were cattle and sheep herders at the time of European colonisation from the 1600s, and who were predominantly located in what is now referred to as the Western Cape. The latter were referred to [pejoratively] as Hottentots by the early European immigrants, and have experienced severe acculturation as a result of colonisation.
2   I prefer not to give the name to protect the area from those who may try to track such sites down.
3   The notion of 'coloured' people can be traced back to the apartheid years during which people were classified on the basis of their skin colour. In reality this was a diverse group of largely 'mixed blood' people, many of whom could trace their ancestry to Khoisan forbears (Marais 1962). It is not a pejorative term in contemporary usage, having been appropriated from the colonisers by people of Khoisan descent.

References

Berglund, A.I.. 1976. *Zulu Thought Patterns and Symbolism*. London: Hurst and Co.

Bernard, P.S. 2008. The fertility goddess of the Zulu: reflections on a calling to Inkosazana's Pool. In *Deep Blue: Critical Reflections on Nature, Religion and Water*, edited by S. Shaw and A. Francis. London and Oakville, CT: Equinox, 49–65.

Bernard, P.S. 2010. Messages from the deep: water divinities, dreams and diviners in Southern Africa. Unpublished Ph.D. thesis in anthropology, Rhodes University, Grahamstown, South Africa.

Bernard, P.S. 2013. 'Living water' in Nguni healing traditions, South Africa. *Worldviews: Global Religions, Culture, and Ecology*, 17: 138–149.

Bryant, A.T. 1949. *The Zulu People: As They Were Before the White Man Came*. Pietermaritzburg: Shuter and Shooter.

Callaway, H. 1970 [1884]. *The Religious System of the AmaZulu*. Cape Town: C. Struik.

Dendle, P. 2006. Cryptozoology in the medieval and modern worlds. *Folklore*, 117 (August): 190–206.

Drewal, H.J. (ed.) 2008. *Sacred Waters: Arts for Mami Wata and Other Divinities in Africa and the Diaspora*. Bloomington and Indianapolis: Indiana University Press.

Eberhart, G. 2002. *Mysterious Creatures: A Guide to Cryptozoology*. 2 vols. Santa Barbara, CA: ABC-CLIO.

Edie, J. 1965. Notes on the philosophical anthropology of William James. In *An Invitation to Phenomenology: Studies in the Philosophy of Experience*, edited by J. Edie. Chicago: Quadrangle Books.

Eliade, M. 1991 [1952]. *Images and Symbols: Studies in Religious Symbolism*. Princeton, NJ: Princeton University Press.

Fabian, J. 2001. *Anthropology with an Attitude: Critical Essays*. Stanford, CA: Stanford University Press.

Goulet, J.G.A. and Miller, B.G. (eds) 2007. *Extraordinary Anthropology: Transformations in the Field*. Lincoln, NE and London: University of Nebraska Press.

Greef, C. and Boshoff, R. 2002. A tale of paradise. *Sunday Times* (South Africa), 24 November.

Hoff, A. 1997. The Water Snake of the Khoekhoen and /Xam. *South African Archaeological Bulletin*, 52: 21–37.

Hoff, A. 2007. *Medicine Experts of the /Xam San: The !Kwa-ka !Gi:ten Who Controlled the Rain and Water*. Cologne: Rüdiger Köppe Verlag.

Jackson, M. 1989. *Paths Toward a Clearing: Radical Empiricism and Ethnographic Enquiry*. Bloomington and Indianapolis: Indiana University Press.

James, W. 1976. *Essays in Radical Empiricism*. Cambridge, MA: Harvard University Press.

Jung, C.G., von Franz, M.L., Henderson, J.L., Jacobi, J. and Jaffé, A. 1964. *Man and His Symbols*. London: Aldus.

Jung, C.G. 1968 (1st edn 1959). *The Collected Works of C.G. Jung. Volume 9*. Edited by H. Read, M. Fordham, G. Adler and W. Mcguire, translated by R.F.C. Hull. London and Henley: Routledge and Kegan Paul.

Kendall, K. 1999. The role of *izangoma* in bringing the Zulu goddess back to her people. *The Drama Review*, 43(2): 94–117.

Kohn, E. 2015. Anthropology of ontologies. *Annual Review of Anthropology*, 44(1): 311–327.

Mansfield, V. 1995. *Synchronicity, Science, and Soul-Making*. Chicago and La Salle, IL: Open Court.

Marais, J.S. 1962. *The Cape Coloured People 1652–1937*. Johannesburg: Witwatersrand University Press.

Morrow, S. and Vokwana, N. 2003. 'Oh! Hurry to the river': *u-Mamlambo* in the Eastern Cape, South Africa. The Eastern Cape: Historical Legacies and New Challenges Conference, East London, South Africa, 27–30 August.

Newton, M. 2005. *Encyclopedia of Cryptozoology: A Global Guide to Hidden Animals and Their Pursuers*. Jefferson, NC: McFarland and Co.

Pienaar, A. 2009. *The Griqua's Apprentice: Ancient Healing Arts of the Karoo*. Roggebaai, South Africa: Umuzi.

Schutz, A. 1974 [1973]. *The Structures of the Life-World*. London: Heinemann Educational.

Sharps, M.J., Newborg, E., Van Arsdall, S., De Ruiter, J., Hayward, B. and Alcanter, B. 2010. Paranormal encounters as eyewitness phenomena: psychological determinants of atypical perceptual interpretations. *Current Psychology*, 29: 320–327.

Soga, J.H. 1930. *The South-Eastern Bantu (Abe-Nguni, AbaMbo, AmaLala)*. Johannesburg: Witwatersrand University Press.

Soga, J.H. 1931. *The AmaXhosa: Life and Customs*. Alice, Eastern Cape, South Africa: Lovedale Press.

Viveiros de Castro, E. 2015. Who is afraid of the ontological wolf? Some comments on an ongoing anthropological debate. *Cambridge Journal of Anthropology*, 33(1): 2–17.

Young, D.E. and Goulet, J.G.. (eds.) 1994. *Being Changed by Cross-Cultural Encounters: The Anthropology of Extraordinary Experience*. Ontario: Broadview Press.

8   Mermaids in Brazil

The (ongoing) creolisation of the water goddesses *Oxum* and *Iemanjá*

Bettina E. Schmidt

Introduction

Mermaids symbolise the ultimate 'other' (Kramer 1987: 213), like humans but not human, like fish but not fish, living in rivers, lakes, waterfalls, springs and in particular the sea. They are usually described as beautiful women who bewitch men into following them into the water and to death. With their beauty and wealth they attract envy and lead people into ruin. Eternalised in many fairytales there is, however, another side to the story of mermaids. During the colonial period in Africa mermaids became associated with Europeans due to their light skin, long straight hair, their wealth and their Otherness (Kramer 1987: 221). Shortly after, in West Africa, they were used to represent local water goddesses who developed into one pan-African deity named *Mami Wata* (also written *Mammywater* or *MamiWata*).[1]

I came across the mermaid as representation of a water goddess during my recent research in Brazil, though not under the name *Mami Wata*. People refer to the water goddesses under their local names, *Iemanjá* and *Oxum*, two deities in the pantheon of the Afro-Brazilian religion Candomblé.[2] Nonetheless, the representation of not only one but two water deities as mermaids in Brazil opens the door to an interesting development corresponding to the decline of ethnic differences between the various local variations of Candomblé within an urban context. We may even observe here the beginning of a new creolisation process that can lead, as is the case for the African *Mami Wata*, to an emergent pan-Brazilian deity. In this chapter I will investigate whether Alex van Stipriaan's interpretation of *Mami Wata* as 'the creolized symbol of the "other", presenting solutions where the familiar traditional gods no longer suffice' (2002: 95) can be applied to the Brazilian context or whether the usage of a mermaid as a symbol for the water goddess is just 'old wine in new skin' (Egonwa 2008). I will look at the potential of a creolised symbol for the transformation of the (local) Afro-Brazilian religions into a unified belief system located within the Brazilian metropolises. Following van Stipriaan's work on the *Watramama* in Surinam, I will further discuss the implication of this transformation from local goddesses into an urban, supra-local deity, a symbol of modernity and globalisation, for Brazil and the Afro-Brazilian religions.

Representation of the African water goddess as mermaid

The core area of *Mami Wata* is the region from Cameroon to Ghana but she is known in many other countries. In a much standardised way she is portrayed as a mermaid with long straight hair and a light skin (generally non-African features – see also Bernard, this volume). Kathleen O'Brian Wicker describes the *Mami Wata* cult as a 'complex transcultural phenomenon composed of elements from widely disparate places and traditions' (2005: 5629). Though I am not interested in extracting the 'authentic' roots of a cultural phenomenon (Hackett 2008: 406), it is important to stress the diversity of the roots of the *Mami Wata* representation because it will be significant later in the discussion of the Brazilian case.[3]

Scholars trace the arrival of European images of mermaids in West Africa back to figureheads on Dutch and other European ships in the pre-colonial era. Early European travellers have been associated with the sea and marine spirits from the fifteenth century onwards and, as a consequence, Africans added 'to their ancient pantheons of water deities a spirit that has come to be known as MamiWata' (Drewal 2002: 193). The earliest documented example of a mermaid is an Afro-Portuguese ivory from Sherbo, Sierra Leone that was brought to Europe in 1743 (Drewal 1988: 103; Kramer 1987: 222). Early in the nineteenth century marine figures became part of water divinity altars, and mermaid-figures developed into a common representation of local water deities (Wicker 2005: 5630). Despite these early usages of the mermaid for local water goddesses, the name *Mami Wata* was probably unknown in Africa before the twentieth century but developed in the Americas. Kramer (1987: 222) mentions early references to the name of *Mami Wata* (as *Waturmamma* or *Watramamma*) in Surinam and Haiti in the mid-eighteenth century, from where the name, together with some iconographic elements, was brought to West Africa via freed slaves (Kramer 1987) or by African sailors (Wicker 2005: 5630). Another author (Wendl 1991) points to the importance of the Kru (an ethnic group in today's Liberia) for the spread of a relatively stan-dardised representation and cult of *Mami Wata* in West Africa. Wendl (1991: 113–116) argues that the Kru became intermediaries for European traders along the coast and initiated the *Mami Wata* cult wherever they went.

Another reason for the creation of a pan-African *Mami Wata* cult, with relatively homogenised iconography and myths, can be traced back to the influence of European images in West and Central Africa. Scholars refer in particular to an early German chromolithograph of a female Indian snake-charmer who performed in the Carl Hagenbeck show in Europe. Posters of the snake-charmer appeared in Africa early in the twentieth century, where it became regarded as a picture of *Mami Wata* (Drewal 1988: 114). Later, large quantities of posters of the snake-charmer were produced in India where the image of the snake-charmer – with more distinctly Asian features – merged with images of Hindu gods. However, any relative standardisation of *Mami Wata* imagery does not mean that *Mami Wata* is a monolithic cult (see

Hunter-Hindrew 2008). The relatively homogenised way of describing and speaking about *Mami Wata* only reflects the popularity of the mermaid goddess. Drewal (1988: 96), who traces the influence of the poster in forty-one cultures in fourteen African countries, gives the following explanation for the popularity:

> Africans determined that there was a direct connection between these Indian images, the beliefs associated with them, and Indians' success in financial matters, just as mermaids and other European icons such as marine sculptures and saints'statues have been linked with wealth and power.
>
> (Drewal 1988: 118)

However, despite the mixture of African, European and Asian elements[4] and the Pidgin name, *Mami Wata* is not a Pidgin phenomenon but an African one, as Drewal (1988: 102) insists. *Mami Wata* has developed into a pan-African water goddess through a process of 'active interpretation, adaptation, and re-creation, not reproduction' (Drewal, quoted in van Stipriaan 2002: 94). Today, *Mami Wata* is well known, even in the West African popular press and tabloids (Bastian 1998: 21). While she still represents local water deities, she has become a popular icon with a special group of artists, the *watistes*, who produce figures and paintings for their clients. One of these artists, Chéri Samba, who has become very successful in the world of international art, warns in many of his paintings against the temptations of *Mami Wata* (van Stipriaan 2002: 96). *Mami Wata* is nowadays often regarded with suspicion, as dangerous, or, as Meyer writes, with regard to Southern Ghanaian Christians, one of the fallen angels who had been driven out of heaven together with Satan (Meyer 2008: 385). As van Stipriaan states, *Mami Wata* embodies all aspects of modernity, both negative and positive:

> She can be unpredictable and aggressive as urban life itself, she can make people rich, or drop them like a hot potato, she relates to individuals rather than the community and her insistence on unconditional loyalty from adherents is interpreted by some as a signal for new rules of sexual contact.
>
> (van Stipriaan 2002: 95)

Mermaids in Brazil

Brazil is the home of several religions derived from African traditions. However, despite the existence of water deities in the pantheon of all Afro-Brazilian religions (or perhaps because of it[5]), there is no pan-Brazil water deity comparable with the *Mami Wata* cult in West Africa. The term *Mami Wata* is also unknown in Brazil. The standardisation process that established the pan-African *Mami Wata* has not taken place in Brazil, nor in other areas of America, as van Stipriaan comments. 'Unlike their West-African sister, the

Afro-Caribbean water mothers seem to have stopped creolization a long time ago and, therefore, might now even be in a state of de-creolization' (van Stipriaan 2003: 327). He argues that the African *Mami Wata* cult flourishes in the post-colonial era of mass communication and increasing mobility and can be seen as a result of psychological and cultural chaos and socio-economic transitions. The American water deities on the other side of the Atlantic 'creolized and gained momentum during the traumatic experiences of slavery when mobility within (let alone between) these slave societies was limited as much as possible. Intra-African creolization, therefore, was confined to the (insular) borders of these individual colonies' (van Stipriaan 2003: 331). Due to limited means of communication, even within the Portuguese colony, Brazil has seen the development of various Afro-Brazilian religions in different locations connected to slavery, all of them the result of a similar creolisation process, but they failed to establish a nationwide system.

Candomblé, today the most well-known Afro-Brazilian religion, is constructed around a pantheon of *orixás*,[6] among them the water goddess *Iemanjá*, who is regarded by some as the mother of all other *orixás* (Silva 2005: 78). *Iemanjá* is derived from the West African Yoruba deity *Yemaya*[7], the goddess of the river Niger. However, in Brazil she occupies the place of *Olocum*, the *orixá* of the sea in Nigeria (Prandi 1991: 130). The transformation from a river goddess to the central salt-water goddess has also entered the mythology. According to legends, *Iemanjá* had to run away from her husband: she left the river and disembarked into the sea where she lived together with her mother *Okun* (= *Olocum*). In Brazil she then became associated with marine spirits and other indigenous water spirits (Silva 2005: 78). Today she is regarded as the goddess of all waters and is especially connected to the Atlantic Ocean. Like the other *orixás* in America, *Iemanjá* also has a Catholic correspondent – in her case *Nossa Senhora* ('Virgin Mary') in different local variations. In Bahia the day of *Iemanjá* is celebrated on the second of February, the day of *Nossa Senhora dos Nevegantes e das Candeias*, while in Rio de Janeiro and São Paulo the festival is celebrated on the eighth of December, the day of *Nossa Senhora da Conceição*. Another important day that is marked with a special ceremony for *Iemanjá* is New Year's Eve. A growing number of people celebrate the New Year with offerings to her at the beach, whether it is in Salvador, in Rio de Janeiro or other places on the Atlantic coast, and also at lakes and rivers in the interior of Brazil. At each place thousands of adherents bring flowers, perfume and other offerings and put them into the water in order to pray for a good new year (Silva 2005: 79).[8]

Iemanjá's transformation from freshwater to salt-water goddess in America might explain the representation of *Iemanjá* as mermaid, popular in particular in Salvador da Bahia in the northeast of Brazil. Salvador is described as the city of *Iemanjá* and many paintings by popular Bahian artists show her as a mermaid, protecting sailors and fishermen. During my recent visit to Salvador I took a photograph of a large statue of *Iemanjá* as a mermaid in front of a house of local fishermen in Salvador, the *Casa de Yemanya*. The mermaid has flowers and the insignia of *Iemanjá* in her hands and is portrayed with long straight hair and light skin. The usual popular representation of *Iemanjá* in Brazil is a woman with long black hair and a blue dress. She is often accompanied by marine animals such as fishes, dolphins or sea-horses. Similar, smaller statues can be bought in little shops all over Brazil that sell paraphernalia to the adherents of Afro-Brazilian religions, including statues of *orixás, Umbanda* spirits and *caboclos*,[9] necklaces in various colours, candles, perfumes, even literature about the religious practices. However, *Iemanjá* is not the only water goddess who is represented as a mermaid.

In 2010, while conducting an empirical study on Afro-Brazilian religions in the metropolis of São Paulo, I encountered another mermaid quite unexpectedly. This time she represented *Oxum*, a freshwater goddess. In the Yoruba pantheon in West Africa, *Oxum* is connected to a river of that name as well as to lakes, springs and waterfalls. Similar to *Mami Wata* (Drewal 1988: 104) she is often portrayed with a mirror in her hand or gazing at her reflection in a mirror. Her Catholic correspondent in Brazil is also the Virgin Mary in different variations, in particular *Nossa Senhora da Conceição, Nossa Senhora das Candeias*, and *Nossa Senhora Aparecida*, the national patron of Brazil (Silva 2005: 78, 94). The representation of Oxum as mermaid was found in a small Candomblé *terreiro* ('house')[10] in São Paulo. The founder of the religious community house is *pai* ('father')[11] Francisco, a Candomblé priest (*babalorixá*[12]) who was initiated to Candomblé in Salvador da Bahia but moved to São Paulo soon afterwards. After meeting him on several occasions where he presented to the Afro-Brazilian community in general, he agreed to a further private meeting. For the interview we went to the office where he conducts oracle readings for members of his community and other clients.[13] On the desk, next to the divination board, I noticed at once the expressive statue of a mermaid in golden colour and decorated with the usual insignia of *Oxum*, the *orixá* of *pai* Francisco. The figure took me by surprise because I did not expect to see *Oxum* as a mermaid. Mermaids are usually connected to a salt-water environment (cf. Bernard, this volume), while as already noted, *Oxum* is a freshwater goddess. *Pai* Francisco was very proud of this beautiful figure and saw no problem in referring to the water goddess *Oxum* as mermaid. As a highly political person he stresses at every possible occasion the African identity as part of Brazil. He demands that the adherents of Afro-Brazilian religions should not only wear African dress during ceremonies but also outside, when shopping in the market or speaking to politicians or representatives of other religions. Only by celebrating African heritage as 'normal' can one overcome the prejudices against Afro-Brazilians (interview on 21 May 2010). However, the pantheon of spirits and deities honoured in his community includes several non-African entities, something that some Candomblé priests reject. *Pai* Francisco argues that as he is Brazilian and lives in Brazil he needs to honour the local spirits of the Brazilian soil (and water), the indigenous spirits. As was customary in the house in Salvador where he became initiated, and to which lineage he therefore belongs, he celebrates once a year a ceremony for *cacique* (term for the 'chief' of an indigenous community) *Pena Branca* ('White Feather'). I also noticed small shelves on the wall of his office with statues of the national patron of Brazil and other entities, though *Oxum* and *Pena Branca* are the main entities. *Pai* Francisco represents a very Brazilian religious mixture and acknowledges the importance of adaptation and syncretism. However, this attitude is not shared by all adherents of Candomblé.

Soon after my interview with *pai* Francisco I mentioned the mermaid to another devotee of *Oxum* (*filho de Oxum*[14]) from a different religious community. He rigorously denied the possibility of representing *Oxum* as a mermaid. Even when I showed him the picture, he dismissed it as wrong. He insisted that mermaid was not an African tradition but a Greek one and should not be used to represent the African goddess *Oxum* (email exchange, June 2010). This son of *Oxum* is a well educated young man who represents a relatively new group of adherents of Afro-Brazilian religions: white, educated and usually urban Brazilians who convert to Afro-Brazilian religions in growing numbers. He belongs to a community in the Candomblé Fon tradition that is regarded as secluded and traditional.

His categorical dismissal is symptomatic of the ongoing debate in Brazil on authenticity and the search for 'true' African traditions. Influenced by Pierre Verger (1902–1996) and other intellectualists and scholars (see Sansi 2007), the Bahian version of Candomblé, emphasising the Yoruba tradition, was for a long time in Brazil considered to have represented Africa in its purest essence. However, in recent decades this emphasis has shifted towards the so-called Bantu tradition within Candomblé that was not 'polluted' by syncretism with Roman Catholicism during the time of slavery, but remained in secrecy an 'authentic' African tradition.[15] This debate about which tradition is more African disregards the evolution of the Afro-Brazilian religions during and after the time of slavery.

When enslaved Africans arrived in Brazil, they continued – despite prohibition – to practise African customs and developed in a relatively unorganised way a *culto africano* ('African cult') that can be seen as a creolised form because it combined the traditions of various African ethnic groups. During the nineteenth century *casas de candomblé* ('candomblé houses') were established which became the birthplace of the tradition combined today under the term Candomblé (or better Candomblés) in order to honour local variations. The Bahian version of Candomblé emphasises the Nago nation which derived from the Yoruba tradition. Candomblé Jêje derived mainly from the Ewe-Fon tradition and Candomblé Angola or Congo from a group of traditions usually labelled Bantu. Another tradition is called *Xangô*, after the Yoruba deity with the same name, hence it resembles Candomblé Nago. And in addition there is *Tambor de Mina*[16] that has a strong influence from Dahomey (today's Benin) and many similarities with the Haitian religion Vodou. Reginaldo Prandi categorises the different forms of Candomblé and the other Afro-Brazilian religions as ethnic religions (2005: 13–14). They all developed in certain areas of Brazil: *Tambor de Mina*, for instance, in the state of Maranhão, *Batuque* in

Rio Grande do Sul and in the Amazonian region, *Macumba*[17] in Rio de Janeiro, and *Xangô* in Recife (Harding 2005: 120) – each region with a strong historical link to slavery (e.g. slave markets or large sugar plantations).

After the final abolition of slavery in 1888, the constitution of the new republic in 1889 declared freedom of religion and abolished Roman Catholicism as the official religion of Brazil. However, Afro-Brazilian religions were still persecuted throughout the twentieth century and Catholicism remained the 'almost official' religion (Mariano 2001: 145, quoted by Oro 2006a: 9), despite the constitutional separation of state and church. Ari Pedro Oro writes that:

> We tend to forget that perhaps the biggest historical victims of religious intolerance, and denial of their religious freedom, have been and continue to be the Afro-Brazilian religions, which together with *kardecism*, throughout decades, have been the target of persecution, given that their ritual practices were seen as acts of fraud, faith healing and charlatanism, either on the part of the press and intellectuals, or on the part of the very Catholic church who, during the '50s, launched a battle against religions which believe in and accept mediums or seers.
>
> (Oro 2006a: 10)

In 1965 it became possible to legalise Afro-Brazilian places of worship by civil registration and for religious communities to apply for tax exempt status as non-profit, charitable institutions (Brown 1986: 3).[18] In particular animal sacrifice, which is an important obligation in all Afro-Brazilian religions and a crucial part of many rituals, was the target of legal prosecution – and is still the target of campaigns against Afro-Brazilian religions, despite all efforts of some outstanding priests and priestesses of Afro-Brazilian religions to increase the visibility and acceptance of their religions. The result is the lack of nationwide institutionalisation, which, as Prandi complains, affects the growth of Afro-Brazilian religions (Prandi 2005: 223–232, quoted in Malandrino 2006: 40). It also supports the persistent identification of Afro-Brazilian traditions with local regions.

However, due to the growing internal migration within Brazil, these local religions migrate, too, especially now that the practice of Afro-Brazilian traditions has been legalised in Brazil. Consequently, one can find different forms of Candomblé in every large city in Brazil. This process increases competition: for membership between different religious houses (*terreiros*), for authority between different priests and priestesses, and about the 'purest' African tradition. Pressure is even enforced by the recent spread of a so-called 'Yoruba tradition' in Brazil. Initiated by Nigerian immigrants who establish and then lead the communities, it is meant to (re)introduce the 'correct' Yoruba tradition and in particular the cult of *ifá*.[19] The consequence of this competition is a growing demarcation between communities and less willingness to cooperate. Even when priests and priestesses attend ceremonies in other *terreiros*, they will always insist, to members of their own community, that their own way to conduct the rituals is the best, the only effective or the 'true' African way. In this situation the reference to mermaids, which have become part of a globalised popular culture, can be seen as 'not African enough'. However, the 'ordinary' adherent is often 'shopping around' and open-minded to new 'offers' on the market of religious traditions. The result of this increasing mobility is an ongoing interaction between different religious communities and different traditions. In this situation the growing presence of the mermaid can offer familiarity in unfamiliar urban settings and can help people 'to cope with great transitions, and with their sense of being uprooted' (van Stipriaan 2003: 330).

Iemanjá, Oxum and *Mami Wata* – the birth of one truly 'creole' icon of the twenty-first century?

Comparing the post-colonial *Mami Wata* cult in West Africa with the Caribbean water deities during the Middle Passage and on the slave plantations, van Stipriaan states that

> the Mothers of Water helped people to find a new individuality and at the same time created a new 'we' in a context in which most people were 'others', and as a reaction to a dominant culture, be it colonial or a (westernised) global culture.
>
> (van Stipriaan 2003: 330)

However, despite many social and political problems, increasing mobility and mass communication, the Caribbean societies did not develop 'enough common ground' to create a pan-Caribbean water goddess. The situation was similar in Brazil, though recently one can notice important changes.

Brazil is confronted with radical transformations, not only in the social and political sectors but also in the religious. Though the majority of Brazilians (64.6 per cent according to the 2010 national census) still declare their belonging to Roman Catholicism, the number is in decline. The national statistics indicate that a growing number of Brazilians belong to one of the numerous Protestant churches (already 22.2 per cent in 2010 with a rising tendency). Both groups together come to nearly 90 per cent of Brazilians. The remaining 10 per cent are divided between spiritists (2 per cent), adherents of an Afro-Brazilian religion (0.3 per cent), agnostics or atheists (8 per cent), and members of another religion such as Judaism, Islam, Hinduism, or Buddhism (Instituto Brasileiro de Geografia e Estatística 2012). These numbers do not cover all adherents of Afro-Brazilian religions because many still avoid being identified with an Afro-Brazilian tradition. Often they will claim to be Roman Catholic or even atheist instead of a member of an Afro-Brazilian community. The term spiritism is also used as an umbrella term to avoid discrimination; hence the number of adherents of Afro-Brazilian religions is probably larger. Nonetheless, even if we take spiritism and Afro-Brazilian religions together, only 3 per cent of Brazilians declared practising a religion in this category in 2010.

However, sociologists are more puzzled by the large scale demographic shift from Roman Catholicism to Protestantism and in particular to Pentecostalism, which also affects Afro-Brazilian religions (Engler 2011). While mainstream, mission-related churches have been relatively tolerant of local traditions, the Pentecostal movement does not usually accommodate them. The Brazilian form of Pentecostalism, labelled neo-Pentecostalism by Ricardo Mariano (1999), is even on a crusade against Afro-Brazilian religions, spiritism and related religions. While the relationship between the neo-Pentecostal groups and the Afro-Brazilian religions was never good, it worsened in 1994/1995 when the *Igreja Universal do Reino de Deus* (Universal Church of the Kingdom of God, short form: UCKG) declared a holy war against them. Despite – or perhaps because of – constant attacks against the Afro-Brazilian religions, many people who have practised *Umbanda* join the Pentecostal churches in great numbers, in particular the UCKG which Oro describes as *macumbeiro* or shaped by 'Macumba'[20] (Oro 2006b). The founder of the UCKG, Edir Macedo, was born into a Catholic family but practised *Umbanda* before he converted to the New Life Pentecostal Church in the 1960s. When he founded his own church in 1977, he included in the liturgy elements from the Afro-Brazilian religions but also the demonisation of Afro-Brazilian spiritual entities already developed in the New Life Pentecostal Church. Deliverance of the demons became the core ritual and means for faith healing.

This attitude against the Afro-Brazilian entities is identical to the way Rosalind Hackett (2008) describes the Pentecostal 'concern' with the mermaid and other marine spirits in Calabar, Nigeria. Referring to the Mountain of Fire and Miracles church in Akoka, founded and led by a British-trained micro-biologist Dr D. K. Olukoya, Hackett explains that marine spirits, and in particular mermaids with their half fish, half human form, are regarded as the most potent, and more dangerous than the earth spirits (the ancestors). Marine spirits seduce people with luxurious items (clothes, jewellery) and expensive habits (such as eating out and dancing), and cause problems such as promiscuity and polygamy. One can even become polluted by a parent or partner worshipping water deities.

> As I was told in hushed tones by one deliverance specialist, so sophisticated are some of these newer 'manifestations' of MamiWata as virtuous church virgins, that they can manipulate through their charismatic charms not just the bodies but also the *minds* of men. He added that this type of spirit is the most dangerous and deceitful, and attacks pastors only.
>
> (Hackett 2008: 410–411)

The Brazilian crusade does not focus on marine spirits alone, as one can see similar tendencies regarding female spirits and women. *Exu* is considered in Candomblé as the messenger between humans and the *orixás* and must be honoured at the beginning of every ritual, in order to gain access to the world of the *orixás*. The Roman Catholic Church, however, associated him early on with the devil. This correspondence was then carried on into *Umbanda*, which has developed a range of different *Exus* as well as a female counterpart, *Pombagira*, who has now become a special target of the UCKG. This focus on female entities (though not exclusively) did not even stop at the national patron of Brazil. In 1996 a pastor of the UCKG kicked the statue of *Nossa Senhora Aparecida* during a service that was broadcast on TV. Though this 'kicking the saint' incident hit back and forced the UCKG, including Macedo, to apologise for this behaviour in public (Almeida 2003: 321; Birman and Lehmann 1999: 150), it shows how deep the aggression against female deities runs in the UCKG. Considering that *Nossa Senhora Aparecida* is regarded as the Catholic representation of the water goddess *Oxum*, we have here another attack against an important water *orixá*.

Nonetheless, does being a target of vicious attacks support the creation of a 'creole icon' as I suggested above? On one side there are well organised and growing Pentecostal institutions that demonise the practices of Afro-Brazilian religions and encourage their members to attack (verbally, symbolically and also sometimes physically) Afro-Brazilian religious communities and adherents wherever possible (see Silva 2007). Due to their political influence they will have an increasing impact on Brazilian society. On the other side we have a still diverse range of Afro-Brazilian religions with two water goddesses, *Iemanjá* and *Oxum*, as part of a larger pantheon of deities derived from African traditions. As I have shown above, there is an older custom of representing *Iemanjá* (who has been transformed during the period of transatlantic slavery into the salt-water goddess) as mermaid, though *Oxum* has similarities with *Mami Wata* too. Nonetheless, representations of *Oxum* as mermaid are not accepted by all adherents. Due to a lack of consistency and institutionalisation, Afro-Brazilian religions fail to attract new members in large numbers and many of their members refuse to acknowledge their faith in public. The result is declining visibility and a lack of political representation of adherents of Afro-Brazilian religions, despite the increasing visibility of Afro-Brazilian culture and identity.[21]

In this situation the water goddess can be a crucial figure. But the question remains whether this symbolises a modern form of the ancient water deities, as noted by Egonwa (2008: 218) about the *Mami Wata* phenomenon in Benin, or whether it is part of a transformation process. In Haiti *Lasirèn*, a *lwa*[22] associated with seduction and wealth is portrayed as mermaid and is regarded as '*Ezili* of the Waters'. Some see her as sister of the two major female *lwas*, *Ezili Freda* and *Ezili Dantò*. However, due to the difficult economical situation, *Lasirèn* today plays an increasingly important role in the Vodou pantheon. Marilyn Houlberg (2008: 565) states that *Lasirèn* echoes *Mami Wata* 'in her duality, her fierceness, and her modernity' and she even reports a growing application of *Mami Wata* elements to *Lasirèn*, though Houlberg interprets this development as recycling of elements of American popular culture into Haitian water spirits (2008: 567). In this situation Egonwa's interpretation of 'old wine in new skin' seems to be appropriate for describing the recent adaptation of *Mami Wata* elements in Haiti.

However, Brazil is in a different situation. Though American popular culture also plays an important role in Brazil, the term *Mami Wata* is still unknown despite an escalating creolisation process. As mentioned, the differences between the local ethnic religions have already started to become blurred in the new environment of the Brazilian metropolis. This development will enforce the creolisation process that 'may also involve an element of unification and innovation featuring the explicit airing of ethnic-cultural differences in rituals' (van Stipriaan 2002: 92). *Orixás* will lose some of their local distinctive identity and acquire new roles, similar to the developments in Surinam. After being marginalised by mass evangelism, Western education and government policies in the twentieth century, *Watramama* and the Surinamese Winti religion have remerged in Western temples of art and culture, as van Stipriaan highlights. Winti has become reappraised 'as a way of life and a form of spirituality relevant to them [the Afro-Surinamese intelligentsia]' (van Stipriaan 2002: 92). Increasing scholarly interest and the support of specialists such as Afro-Surinamese migrants in the Netherlands have moved *Watramama* from the spectrum of a folk religion (Winti) into 'the upper social echelons in Suriname and the Netherlands' (2002: 92). The same is happening in Brazil where Afro-Brazilian culture – in particular elements of the Bahian Candomblé – has become part of an elite vision of Brazil. The interest shown by the intelligentsia in Afro-Brazilian heritage seems to exaggerate the reality of the marginalised Afro-Brazilian religions. The *orixás* have moved from religious confinement to popular culture and even elite culture, with its museums, art galleries, theatres and literature. One can find sculptures and paintings of *orixás* in the galleries of urban Brazil and even in European and North American museums and art galleries. *Orixás* are now part of the global culture. In particular the water goddesses have the potential to offer an alternative in a globalised world without losing their local identity.

Notes

1 The name *Mami Wata* derives from Pidgin English for 'water as mother' and is even used today in francophone countries in Africa.
2 Throughout the chapter I will use Brazilian spellings for Brazilian religions and Brazilian deities and names.
3 As the focus of this chapter is on Brazil, I will skip over many details though I do not want to create the impression that the development of a *Mami Wata* cult in West Africa was a homogeneous process, quite the opposite. It was the outcome of a long and complex interaction between conflicting and even confusing aspects which fall beyond the scope of the current project.
4 Van Stipriaan (2003: 326) mentions that roots might be traced in the Muslim world, too.
5 According to Derwal (2008: 3), *Mami Wata* is also absent among the Yoruba. He guesses that the reason for this absence is the existence of an ancient and well-defined pantheon of water goddesses.
6 Brazilian term for African deities, derived from the Yoruba language in Nigeria.

7 In order to distinguish between the Brazilian deity and the West African one, I
   have chosen the English spelling for the West African entities and the Portuguese
   spelling for the Brazilian one.
8 Similar ceremonies are even held in Coney Island, New York, though mostly by
   adherents of the Cuban *orisha* religion.
9 Group of indigenous spirits.
10 Brazilian term for the house or site of a religious community; used for Candomblé
   as well as *Umbanda* communities.
11 *Pai* is a short form of *pai de santo* (father of the saint), the Brazilian term for a
   priest, and used in all Afro-Brazilian religions; the female equivalent is *mãe de
   santo* (mother of the saint). The short forms are used in addressing priests and
   priestesses as a form of respect.
12 Priest of an Afro-Brazilian religion (derived from the West African Yoruba language);
   the female equivalent is *ialorixá*.
13 The Brazilian expression for this divination technique is *jogo de búzios*. It is conducted
   with small kauri shells (called *cauris* or *búzios*). During the long initiation process the
   priest or priestess learns how to interpret the way the shells fall on the board.
14 Son of *Oxum*: Brazilian expression for someone initiated into the cult of *Oxum*;
   the female equivalent is *filha de Oxum,* daughter of *Oxum*.
15 The same development can be observed in the USA and the Caribbean, particularly
   among adherents of Afro-Cuban religions. For nearly 100 years the Cuban *orisha*-
   religion (formerly called *Santería*) was considered the true Afro-Cuban tradition,
   but recently the Palo Monte religion – with a stronger emphasis on Bantu instead
   of Yoruba – became praised as 'more African' than the *orisha* religion.
16 The term *Mina* is a reference to the Portuguese fort São Jorge da Mina in West Africa
   where many slaves were imprisoned before transported to Brazil (Silva 2005: 83).
17 Brown (1986: 2) states that people referred to Afro-Brazilian religious groups using
   the generic term *Macumba*, with a pejorative designation; even today it is often
   used with a negative connotation.
18 See Winant (1992) for an overview of the changes of race politics that have initiated the *abertura* (the democratic opening towards Afro-Brazilian cultures and
   religions) in the 1970s.
19 *Ifá* is regarded by some as the most sophisticated cult of the Yoruba tradition;
   priests are traditionally only men though recently it became known that women
   have been initiated into *ifá*, too. However, the Nigerian priests in Brazil are by and
   large *babalorixás* and not ifá-priests (called *babalowas*).
20 Macumba can be described as a form of Umbanda, practised predominately in
   Rio de Janeiro.
21 During the last decade the Brazilian government has successfully improved race relations in Brazilian society: for instance, by supporting black students' access to higher
   education, including references to African heritage in school textbooks and establishing
   cultural centres and museums that present African culture as part of Brazil.
22 *Kreyòl* is the term for deities in the Haitian Vodou pantheon. The term *Lasirèn*
   derives from the French *la sirène* ('the siren').

References

Almeida, R.de. 2003. A Guerra dos Possessões. In *Igreja Universal do Reino de Deus:
   Os Novos Conquestadores da Fé*, edited by A.P. Oro, A. Corten and J.-P. Dozon.
   São Paulo: Paulinas, 321–342.
Bastian, M.L. 1998. MamiWata, Mr White, and the Sirens off Bar Beach: Spirits and
   Dangerous Consumption in the Nigerian Popular Press. In *Afrika und das Anderes:*

Alterität und Innovation, edited by H. Schmidt and A. Wirz. Münster: Lit Verlag, 21–31.

Birman, P. and Lehmann, D. 1999. Religion and the Media in a Battle for Ideological Hegemony: The Universal Church of the Kingdom of God and TV Globo in Brazil. *Bulletin of Latin American Research*, 18(2), 145–164.

Brown, D. 1986. *Umbanda: Religion and Politics in Urban Brazil*. Ann Arbor: University of Michigan Research Press.

Drewal, H.J. 1988. Interpretation, Invention, and Re-presentation in the Worship of MamiWata. *Journal of Folklore Research*, 25(1/2), 101–139.

Drewal, H.J. 2002. MamiWata and Santa Marta: Imag(in)ing Selves and Others in Africa and the Americas. In *Images and Empires: Visuality in Colonial and Post-colonial Africa*, edited by P.S. Landau and D.D. Kaspin. Berkeley: University of California Press, 193–211.

Egonwa, O.D. 2008. The Mami-Wata Phenomenon: 'Old Wine in New Skin'. In *Sacred Waters: Arts for MamiWata and other Water Divinities in Africa and the Diaspora*, edited by H.J. Drewal. Bloomington: Indiana University Press, 217–227.

Engler, S. 2011. Other Religions as Social problem: The Universal Church of the Kingdom of God and Afro-Brazilian Traditions. In *Religion and Social Problems*, edited by T. Hjelm. New York: Routledge, 213–224.

Hackett, R.I.J. 2008. Mermaids and End-Time Jezebels: New Tales from Old Calabar. In *Sacred Waters: Arts for MamiWata and other Water Divinities in Africa and the Diaspora*, edited by H.J. Drewal. Bloomington: Indiana University Press, 405–412.

Harding, R.E. 2005. Afro-Brazilian Religions. In *Encyclopedia of Religion*. Vol. 1. 2nd edn. Edited by L. Jones. Farmington Hills, MI: Thomson Gale, 119–125.

Houlberg, M. 2008. Arts for the Water Spirits of Haitian Vodou. In *Sacred Waters: Arts for MamiWata and other Water Divinities in Africa and the Diaspora*, edited by H.J. Drewal. Bloomington: Indiana University Press, 560–567.

Hunter-Hindrew, V. (Mama Zogbé). 2008. MamiWata – 'It's in the Blood': A Personal Journal of Ancestral Resurrection in the Aftermath of Slavery. In *Sacred Waters: Arts for MamiWata and other Water Divinities in Africa and the Diaspora*, edited by H.J. Drewal. Bloomington: Indiana University Press, 578–591.

Instituto Brasileiro de Geografia e Estatística. 2012. *Censo Demográfico 2010: Características gerais da População, Religião e Pessoas com Deficiência*. Available at: www.ibge.gov.br/home/estatistica/populacao/censo2010/caracteristicas_religiao_defi ciencia/default_caracteristicas_religiao_deficiencia.shtm (accessed 24/8/2012).

Kramer, F. 1987. *Der rote Fes: Über Besessenheit und Kunst in Afrika*. Frankfurt am Main: Athenäum.

Malandrino, B.C. 2006. *Umbanda: Mudanças e Permanencies. Uma Análise simbólica*. São Paulo: Ed. PUC-SP.

Mariano, R. 1999. *Neopentecostais: Sociologia do Novo Pentecostalismo no Brasil*. São Paulo: Ed. Loyola.

Meyer, B. 2008. Mami Water as a Christian Demon: The Eroticism of Forbidden Pleasures in Southern Ghana. In *Sacred Waters: Arts for MamiWata and other Water Divinities in Africa and the Diaspora*, edited by H.J. Drewal. Bloomington: Indiana University Press, 382–398.

Oro, A.P. 2006a. The Sacrifice of Animals in Afro-Brazilian Religions: Analysis of a Recent Controversy In the Brazilian State of Rio Grande do Sul. *Religião e Sociedade*, 1, 1–14. (Translation of the Portuguese article in *Religião e Sociedade*, 25(2), 2005).

Oro, A.P. 2006b. O Neopentecostalismo 'Macumbeiro'. In *Orixás e Espíritos: O Debate Interdisciplinarna Pequisa Contemporânea*, edited by A.C. Isaia. Uberlândia: EDUFU, 115–128.

Prandi, R. 1991. *Os Candomblés de São Paulo: A Velhamagiana Metrópole Nova*. São Paulo: Ed. Hucitec.

Prandi, R. 2005. *Segredos Guardados: Orixásna Alma Brasileira*. São Paulo: Companhia dass Letras.

Sansi, R. 2007. *Fetishes and Monuments: Afro-Brazilian Art and Culture in the 20th Century*. London: Berghahn Books.

Silva, V.G.da. 2005 [1991]. *Candomblé e Umbanda: Caminhos da Devoção Brasileira*. 2nd edn. São Paulo: Ed. Ática.

Silva, V.G.da. 2007. Neopentecostalismo e Religiões Afro-Brasileiras: Significados do Ataqueaos Símbolos da Herança Religiosa Africana no Brasil Contemporâneo. *Mana*, 13(1), 207–236.

Van Stipriaan, A. 2002. Creolization and the Lessons of a Water Goddess in the Black Atlantic. In *Multiculturalismo, Poderes e Etnicidadesna Africa Subsariana* [Multiculturalism, Power and Ethnicities in Sub-Saharan Africa]. Porto, Portugal: Centro de Estudos Africanos, 83–103.

Van Stipriaan, A. 2003. Watramama/Mami Wata: Three Centuries of Creolization of a Water Spirit in West Africa, Suriname and Europe. In *A Pepper-Pot of Cultures: Aspects of Creolization in the Caribbean*, edited by G. Collier and U. Fleischmann (*Matatu*, 27–28). Amsterdam: Editions Rodopi, 323–337.

Wendl, T. 1991. *MamiWata, oder ein Kulturzwischen den Kulturen*. Münster: Lit Verlag.

Wicker, K.O. 2005. MamiWata. In *Encyclopedia of Religion*. Vol. 1. 2nd edn. Edited by L. Jones. Farmington Hills, MI: Thomson Gale, 5629–5631.

Winant, H. 1992. Rethinking Race in Brazil. *Journal of Latin American Studies*, 24, 173–192.

9 Ganka: trickster or endangered species?

An anthropologist's role in preventing the extinction of the New Jersey sea monster[1]

Tanya J. King

Introduction

Deep in the waters off the east coast of New Jersey, USA,[2] lives a creature called the ganka. Ask almost any commercial shark fisher who lives in a coastal town abutting the fishing grounds and they will describe the morphology and behaviour of the animal in great detail. Gankas look like a cross between a cuttlefish and an octopus. Indeed, ganka young are indistinguishable from cuttlefish young, and occasionally surface in the gillnets used to hunt shark. As gankas approach maturity, however, they grow evasive and somewhat mischievous. They also grow short, sharp, teeth. Some fishermen sport nasty scars from ganka bites. Gankas possess great dexterity, their long tentacles enabling them to move through water, and even onto islands and boats, with ease. Though not restricted to salt water, they prefer brine, never venturing too close to the mainland so as to avoid the abundant concentrations of fresh water flowing into the ocean from rivers. The ganka's unusually resilient digestive system allows them to eat a range of non-organic substances including plastic, steel and nylon. This possibly indicates a physiological need for roughage, but also reflects their curious natures and hearty appetites.

Gankas are a pest to fishermen. When boats are laying off the gear (waiting for gillnets to ensnare passing shark), and the crew are sleeping, particularly at night, mature gankas have been known to climb up the anchor rope and onto the deck of the boat. Once aboard, according to experienced fishermen, they will 'eat anything that's not tied down', including ropes, floats and gaffs (sticks with a hooked or barbed point). They have a particular penchant for gumboots. In order to thwart the pillaging, someone must stay awake, on 'ganka watch'. Usually – indeed, without exception – the person who goes on first watch is an inexperienced, uninitiated, 'green' deckhand.

The ambiguous ganka

Gankas are ambiguous creatures; they transgress the boundary between myth and reality. They also blur other boundaries that fishermen recognise. Half in and half out of the water, the boundary that shark boats – and shark fishermen – straddle is important to fishermen both literally and figuratively. The surface of the water acts as a point of inversion, modifying meaning as things pass in and out of the sea. It is the very boundary that both separates and connects that which is above and below the water. For example, below the water is potential economic benefit, while it is above the water that this potential may be realised. Above the water a man is alive; below he is dead. When the creatures that live in the water cross into the air they, too, tend to quickly pass from life to death.

Gankas, however, are not bound by the interface of the sea surface, upon which shark boats are precariously perched, above which men can breathe and beneath which they sink into death. Gankas are at home on land and in the sea. They eat the 'inedible'. They are also almost entirely unknown to non-fishers who spend their lives on land. Gankas move easily between spaces that are connected and yet which seem mutually exclusive. It is precisely their liminal nature that, on the one hand, allows them to move across boundaries and, on the other, connects the domains that they move among. As with Icelandic folk-lore, 'anomalous water-beings ... stress the boundary between land and sea' (Pálsson 1991: 97), by moving between the two.[3]

Because of the ganka's ambiguity, its mischievousness, and its role in the social terrain of a New Jersey fishing fleet, it may be considered a 'trickster'. 'Trickster' is the name given to the creature, often an animal or human-animal hybrid, which appears in the mythological stories of numerous, often polytheistic (Hyde 1998: 9–10) cultures. Stories of tricksters, mostly from North America (Boas 1898; Brinton 1896; Jung 1956; Radin 1913; 1956; Ricketts 1966; Teit 1898, cf. Burnham 1998) and Africa (Evans-Pritchard 1967; Guenther 2002; Pelton 1980; Wescott 1962), have been widely documented and analysed since the 1860s (Doty and Hynes 1993: 13). Many are complex, and involve the trickster as an outsider or someone on the borders of society entering into a lengthy exchange of promises, goods and services, with a range of characters at various levels of power and authority within the community. Sometimes the trickster is a fool, who unsuccessfully tries to outsmart his[4] opponents, or who behaves selfishly and deceptively only to find that his actions have unexpected creative outcomes for humans. Often, the tales are very funny. Tricksters star in tales in which they are destructive – mischievous, insatiably hungry and lustful – and those in which they are creative – life-givers, saviours and cultural 'bricoleurs' (Hynes 1993: 42–45). Tricksters also occupy an ambiguous space. They are often both destructive *and* creative. Judging from the analyses of trickster tales, they routinely serve to illustrate important social values and interpersonal roles, articulate tensions, or highlight inequities in the cultures in which they appear. As theriomorphic figures, or anthropomorphised animals who ignore the boundaries of the natural, the social and the spiritual, tricksters are cosmological artefacts of the most intriguing and incomprehensible kind.

Numerous attempts have been made to capture a definitive account of the 'trickster' (Hynes and Doty 1993; Hyde 1998; Pelton 1980), and some of last century's most talented thinkers have grappled with him (Jung 1956; Koepping 1985; Lévi-Strauss 1963: 202–228; Turner 1969; Douglas 1968: 372–374; Evans-Pritchard 1967), many emphasising his underlying binary composition. The trickster remains, however, a slippery beast. As Pelton noted, over thirty years ago:

> The logic shaping the trickster has been a rock upon which many good scholars have foundered. Most attempts to interpret the tricksters and their myths leave the figure himself curiously untouched; the many angles of approach which interpreters have used have not come to grips with his ambiguity.
>
> (1980: 7)

It is perhaps this ambiguity which is the most common trait of the trickster, or the manner in which his stories, and the analyses of his stories, throw into question the basic social tenets of the society in which he dwells. It is perhaps also that each trickster must be considered in his own social context, which frustrates a common definition. If tricksters are jokers, we should note the words of Mary Douglas: 'a joke cannot be perceived unless it corresponds to the form of the social experience' (1968: 368).

Rather than attempt to fit the ganka into a model definition of the trickster (Hynes 1993) or to reshape the model itself according to ganka qualities, in this chapter I sketch the 'form of the social experience' (Douglas 1968: 368), or the context in which the ganka dwells. By 'context', I allude to the entire gamut of social and physical components of the life-world of fishers; these cannot, in practice, be separated.[5] People, places, productive operations, environmental and economic realities, cultural, corporeal and oceanographic factors are all features of the ganka's habitat. In keeping with this holistic description, this chapter is written with the contemporary political situation of those in the New Jersey shark fishing industry in mind. The situation is one in which the life-world of the fishers – their productive demands, physical boundaries, and social relationships – is changing in response to adjustments to management structures. As such, the habitat of the ganka is changing. It is upon the point of 'ganka habitat conservation' that this chapter concludes.

I progress to this point about ganka habitat conservation in two steps. First, I describe the physical and social habitat of the ganka, and the function of the ganka as a creature that (1) helps to direct the deckie's attention to the salient elements in the environment and, (2) helps weave the deckie into the social fabric of the crew. Second, I describe how I came to know about the ganka, which was via the words of two professional deckies, Byron and Shaun, who have an intimate knowledge of East Coast waters and the creatures that dwell there. I discuss the implications of my knowledge, both for the community in which the creature dwells and for cryptozoologists interested in the ganka's dissection. My point is to highlight the responsibility of scholars to promote the preservation of such creatures, not merely in academia but in their natural habitats.

Where the ganka dwell

The New Jersey fishing community on which this chapter focuses does not correspond to a particular town, or place, or even to a particular fishery. During the most intense period of my fieldwork, during a sabbatical at Rutgers University in 2010, much of the industry was made up of small-time 'lifestyle' fishermen (Minnegal et al. 2003; see also Ota and Just 2008: 305), with a strong sense of personal attachment and belonging to the fishing industry as a whole. In 2010 there were approximately 130 shark licences in the fishery, though some were latent, and many operations comprised multiple licences (shark and, say, sea bass or porgies). Of those who hunted shark I knew no more than forty. While many fishers professed to favour hunting particular species, and more senior fishers (owners and skippers) tended to focus on the species for which they had skill and licences or quota, most fishers' life-histories involved working in a range of different fisheries, including shark. The community encompasses those who venture out into New Jersey waters, primarily those who search for shark, or those who have done so in the past.

Though there are certain areas that are targeted more often than others, there are no specific 'grounds' towards which fishers uniformly direct their vessels when embarking on a shark trip. Rather, boats from different ports, including Belford, Cape May/Wildwood and Barnegat Light all have access to the same stretch of water. The crews of these boats all encounter similar conditions and experiences, and the vast majority have fished from multiple ports and vessels. Most fishermen can identify New Jersey boats from an impressive distance, and can provide a history of the owners, crew, and changes in species targeted over the years. While at sea, many skippers conduct radio or phone relationships with fishers from other ports, some of whom they have never physically met. Thus, the community of interest here is without well-defined geographic limits, without a common residency, and lacking in regular face-to-face contact among members. The community of New Jersey shark fishermen creates and maintains links through a demonstration of common experience that is manifest as gossip and storytelling and in the cooperative act of hunting shark.

How to catch shark

Shark fishers take to sea in small, low draught, boats – 10–20 metres – with a skipper and, ideally, two deckhands. When travelling, the nets are kept wound around a 'spool', which resembles a large cotton reel, on the deck of the boat. Once a fishing spot has been determined the nets are mechanically 'shot' into the water and left to 'soak' on the bottom of the ocean for six to eight hours.

It is during this rest period that the crew can have some time to themselves. This is a time for eating and other bodily functions, watching television, reading, repairs, chatting to other crew and, finally, sleeping. It is a time when the motors are turned off and the noise of the ocean closes in around the boat. Particularly if it is at night, it may also be the time when someone is required to go on ganka watch.

Lead weights and small buoys tied to either side of the net position it in such a way that it sits vertical in the water column. The 'gear' should rest so that the maximum area is netted. Because fishers cannot see their nets in operation, a great deal of thinking, speculation, adjustment and readjustment takes place in order to maximise the area of water netted, with the slightest advantage being judged worthy of years of tinkering. Details of how one configures their leads, buoys and other technical features is important information which is meted out, carefully guarded and seriously debated among some fishers, though many stress the rule of thumb; 'wet nets catch fish'. If the nets have been shot in the right location, at the best angle according to the tides and predicted shark movement patterns, migrating shark will swim into the nets and, unable to move forward, an activity that enables them to breathe, they die.

The physiological aspect of the shark most relevant to its capture, and to the socialisation of shark fishers at sea, is that the shark do not have a pocket of air to regulate their buoyancy; the 'swim bladder'. In other fish species, sound waves from electronic fish-finding technology detect this swim bladder and alert the watchful skipper to the presence of fish below. Lacking a swim bladder, shark are undetectable using such technology and fishers must rely on other signs to locate fish. Like fishers in other parts of the world (McGoodwin 1979; Pálsson 1994), those from New Jersey command an ability to 'read' the 'text' (Moore 1986) of the environment which reflects the intimacy of their productive relationship with their natural and social world.

Many of the deckhands officially employed during the period of my fieldwork were in their late teens or early twenties. Most who skippered or owned boats had begun fishing as teenagers, on the boats of their fathers, uncles or family friends. Fishers have, in the past, been socialised into the seascape as young (usually) men who, in the land-based community, have been in the process of trying to establish themselves as autonomous individuals separate from their male kin (King 2007). The skills to *be* a fisherman, however, involve learning to be part of a highly coordinated team in which the role of the novice is subsumed under that of everyone else. Learning this world – how to recognise its features and respond to its cues – is as important as learning to coil rope, sharpen knives and recognise changes in ocean swells. Indeed, the two are inextricable, and learning to be a fisher involves becoming holistically skilled; the ganka emerges as an important part of the education of deckies.

In a manner akin to that described by Davis (2009) in relation to Polynesian seafarers, New Jersey fishers gather multiple threads of information and spin them into an interactive map which enables them to navigate the seascape with greater proficiency than had they relied on navigational charts, log-books, weather, fish-behaviour, gossip or memory alone. They splice these strands of knowledge, observation and intuition together, fashioning a stable raft upon which to proceed. Some material for the platform is gleaned from the natural world, both that at sea and on land. Tides, the kind of bottom being fished or the type of tree which is currently flowering on land, will provide hints relevant to the search for shark. The first function of the ganka is to direct the attention of the novice deckhand to those details in the environment that can, indeed, be 'read' as part of the productive endeavour.

Professional deckhand Macka explained that when readying to initiate a green deckie, one should point out certain environmental phenomena and comment, with the appearance of knowledge and conviction, that the conditions are just right for gankas. He imitated the ruse – complete with pointing gestures and a furrowing of the brow – with deadpan solemnity: 'You know, you say, "Oh, there's a Southerly, and the water's lapping at the boat a certain way and this bird, or that bird, is flying over: take care lad, it's ganka weather"'. The green deckie may wait in vain for a curious ganka to board the boat and devour his gumboots, even in the most suitable of conditions. However, games like the one Macka described compel the deckhand who may once have per-ceived a rigid distinction between, say, birds and fish, or may have failed to recognise subtle differences in wind direction, to appreciate alternative boundaries and connections and features of salience in the world (Ingold 2001: 113).

The most reliable indicator of fish in the water is fish on the deck. If even a small part of a shoal is detected then the rest can be hunted. This means that once the gillnets begin to be hauled, the skipper is accessing vital information regarding the location of shark. Are there shark or other species in the region? Which direction are they headed (which he can tell by determining from which direction the fish were meshed)? Have they been eating recently? How big is the shoal? If a shoal is deemed to be heading south, the skipper is under pressure to finish hauling and get the nets back in the water in a southerly direction before the fish change direction or somebody else catches them. At this point, time is of the essence and an efficient crew can get their nets back in the water faster than a less coordinated crew. Whether or not their speed has a calculable influence on the overall catch of the boat is less relevant than the fact that the cooperative efficiency of the crew is *understood by fishermen* to have a significant impact on the success of the boat. The ganka contributes to the social dynamic in which deckies move quickly and safely at the command of their skipper. Rather than being a relationship of pure hierarchy, the bond between a skipper and their deckies is one of interdependence (Dumont 1980), that I have elsewhere termed 'prosthetic' (King 2007). Deckies resemble literal extensions of the skipper; living prostheses. While deckies relate to nets, knives and fish with their bodies, many skippers engage with these items in a way that is mediated through the bodies of their deckies. Via their control over the bodies of their deckies, the physical capacities of a skipper are extended, multiplied and strengthened. In a well run fishing operation deckies move around the deck of a boat intuitively, and they relate to the tools of their trade – the deck, fish, knives, cigarettes, coffee-cups – seamlessly. 'The actions of others, contextualised within a shared world of understanding, inform their own' (King 2007: 545).

Part of working and moving as part of a coordinated team is to understand one's relationship with the other men on the boat, and one's role in the system of collective action. Deckies are not simply lower down the 'pecking order' than skippers. However, there is a sense in which the directive to submit one's body to the will of another is to embody an inferior role to the one doing the directing. Acquiescing to such integration requires some training and cajoling. Part of this 'education' often involves an encounter with a ganka, which initiates the deckie into the team, albeit as a minor, peripheral or prosthetic player.

As an anthropologist and a certified land-lubber, I am neither at home on the sea nor in the community of shark fishers, and I am certainly not one traditionally positioned to witness the ganka. Rather, my glimpse of the ganka was largely a result of persistence, luck and alcohol.

The night I encountered the ganka ...

I first learned of the ganka during an evening of heavy drinking at the home of Byron and Shaun, Belford deckies, and brothers. I was there with my friend, Fred, who skippered the *Pamela Joy*, upon which Byron worked. While I had fished with Shaun several times before, I had never taken to sea on the *PJ*.

Byron was a professional deckhand (not on a career path to become a skipper) of around forty years of age. Though Byron had a partner, she lived elsewhere with her children, and his home had only the hints of a 'woman's touch'. Byron and Shaun's decorating style reflected their lives at sea, which had begun as teenage deckhands. Along with nautically-themed household items there were shells, rocks, bones and other items of curiosity which had emerged from the depths, been dried out and displayed. Most notable were the maritime books and magazines which spilled out over the ply-board bookcases and onto piles on the floor. The titles indicated a broad interest in anything maritime. Included in the collection were books on boat building, crustacean biology, archaeological accounts of excavated coastal caves, magazines specialising in boat sales,[6] biographies of famous mariners and fictional tales set at sea.

During serious discussions about the environmental principles under-pinning government fisheries management policies, both men could hold their own. Generally, however, their knowledge was expressed in the witty, common-sense, larrikin-style of many fishers. For example, one of their favourite sayings was 'Fuck the whales: save the plankton!' While obviously good for a laugh, the statement was a serious suggestion that the most viable course of conservationist action in a multi-species, pelagic fishery was considering food webs rather than individual species management.

Their knowledge did not simply come from books, but had been generated over lifetimes of observation and experience. During fishing trips, Shaun would keep a small bucket of 'living curios' – insects, 'strange' weeds, plankton (or so he assured me!) – which he observed over the course of the voyage. I had seen other deckhands do this, and was intrigued by the intense curiosity these men showed for the smallest creature they found, or for the most subtle environmental variation they perceived. During my fieldwork, almost everything was new to me – strange and wonderful and noteworthy – and I could only imagine the depth of familiarity these men had with their environment.

The first time I went shark fishing, Shaun had taught me a lot about how to direct my attention to the salient items in the world around me (Ingold 1992; Pálsson 1991; 1994), a style of attention in which every aspect of the world was interrelated, rather than being an inventory of discrete things, ideas and patterns. Along with the other men on the boat, Shaun would point out islands, changes in the quality of the surface of the water, in wind patterns or fish behaviour, and what this might mean for us, for the fish, or for other species. I was often overwhelmed by the effort of recognising delineations and connections in the world that I had not previously perceived; this perspective was second nature to seafarers like Shaun. For me, that first, eleven-day trip was a harrowing and rewarding experience, during which time I struggled to acclimatise myself to the living quarters, the elements, the social dynamics, the constant motion and the seemingly endless, involuntary emptying of my stomach. Shaun, on the other hand, like his brother, was utterly at home on a fishing boat, wedged in behind the fold-out table of a tiny, cigarette-smoke filled galley with two other fish-gut spattered men, and a 'green' (in both senses of the word) anthropologist.

On the night of the ganka, sprawled around Byron and Shaun's kitchen table, drinking and laughing until we gasped for air, the social tensions which the three men understood implicitly as part of their life-world were slightly relaxed, allowing a glimpse into the fabrication of the bonds that tie fishers together. It was an unusual night for a number of reasons, not merely because of what I learned, or because we were particularly drunk. I had never been to Byron's house even though he lived virtually across the street from me. Further, I had the impression that Fred was not a regular visitor either; any drinking which took place among the men tended to occur at a local bar. In keeping with the relationships between skippers and deckies, in which the former are responsible for the latter, while the latter are deferential to the former (King 2007), this 'hosting' of a skipper by his deckhand felt out of place. There was something about the 'tone' of that evening in which the roles of the men present were slightly askew. There was an openness to their interactions, an intimacy, a subtle levelling of the hierarchy between the skipper and the deckies. Whereas deckies often fade into the background of a social scenario and skippers do most of the talking (King 2007: 548), on this evening Byron and Shaun held the floor.

I could offer a series of hazily recalled anecdotes that would do no justice to the story-tellers or their tales. These stories would include accounts of novice deckhands, their young faces contorted with poorly-disguised fear as they prepare for their first night on ganka watch, as well as the postscript in which they finally realise the ruse and emerge as initiated fishermen. The stories would also include the peculiarities favoured by wily skippers and professional deckies as they prepared the boat – and the green deckie – for a ganka visit. One might reveal a ganka bite scar, collude with others on the boat in telling hair-raising 'near-miss' stories, point out ganka breeding grounds on the map, or give elaborate explanations as to why the deckie may never have heard of such a creature. In some cases the deckie might go on watch for several nights in a row, eventually complaining that someone else should take a turn. In others, the skipper might stumble across a bleary-eyed boy in the wheelhouse of the boat and, upon hearing he was on ganka watch, chuckle quietly, and return to his bunk. But I can only provide these first few glimpses, partly because my recollections are poor, and partly for reasons outlined in the next, and final, section of discussion.

Tanya: rubber-neck[7]

By the time I first heard of the ganka, I had been on several commercial fishing trips, both in New Jersey and abroad. Though these had been merely 'tucker trips' (during which I tried to work, in exchange for food and board), I was no longer derided as someone on the edges of the industry who had 'never been, never seen and never done'. Given my lack of visceral adaptability to the motion of a boat, and the ongoing political, managerial and economic turmoil in which the fishery was embroiled, I had been spending much of my time at governance meetings both locally and interstate, and talking with fishers about the seemingly continual transformations these meetings concerned. As a result, I felt like I had a good understanding of both what happened at sea as well as the political context in which the industry attempted to operate.

I had also reached the point in my fieldwork where I did not feel like I was 'doing fieldwork'. I had friends in the community about whom I cared very much, I worked weekend shifts at a local bar, I had traded in my 'city' boyfriend for a local timber worker and I attended weekend football games. It had become easier to converse with locals about local issues – fishing related and otherwise – than to make conversation with my academic colleagues. Though my place in the New Jersey fishing community was always mediated by my role as an anthropologist, I had become accustomed to my place. However, I was never *really* eligible for incorporation into the community of fishers – I was, like other anthropologists, merely masquerading as 'one of them' – and this was why I did not encounter the ganka in the flesh, on the deck of a boat. Of course, as with other anthropologists, I was often convinced by my own performance. However, the events of that night at Byron and Shaun's, and my 'discovery' of the ganka, forced me to reflect upon my place in the fishing community more critically.

Initially, I was struck by my own sense of disappointment at having been spared the humiliation of being tricked into staying up all night on watch for a wily and fabulous beastie. Despite having professed to being interested in fishing 'culture' I had not been made privy to this 'anthropological gold' which was the ganka. In the coming days I chastised myself for not 'fitting in' enough to be deemed eligible for initiation, but then I regrouped and determined to find out as much about the ganka as I could. How wide did it roam? Did the morphology change as the ganka migrated? Was it restricted to shark boats? How common was the ruse? I would conduct a rigorous investigation of this creature, dragging the depths of the East Coast to uncover every last hiding place of the ganka.

However, when I came to actually question fishermen about the ganka I was struck by another unsettling feeling. I saw my friend Ryan the day after my 'discovery' and when he asked what I had done to earn such an almighty hangover I told him only the basics. I felt like I had possession of something that was not mine. Whereas I felt comfortable asking Ryan almost anything about the fishing industry, including trade secrets and his personal finances, this was different. I did not feel eligible to 'witness' the ganka.

A couple of days later I found myself on a busy Barnegat Light wharf. Though surrounded by bored deckies, most of whom could be cajoled into chatting, I found that I could not bring myself to ask even the more experienced of them about their encounters with the ganka. The risk I feared was that by discussing the ganka I would destroy its secret and thus its social efficacy. I tried inquiring about the creature from the perspective that its existence was a given. In the course of conversation I asked Mike, a quick and witty skipper whom I assumed had duped many deckies over the years, why gankas did not come onto the mainland. With only the slightest hesitation Mike explained that the freshwater run-off from the mainland was too much for the salt-water creature and that they tended to stay a considerable distance off-shore. However, Mike did not take the chance to elaborate on the ruse. I feigned satisfaction with his concocted answer, and was immediately consumed by guilt for my own deceit.

I never did ask a deckhand about the ganka, though over the next couple of months I enquired of several other skippers, who all reported similar encounters involving green deckies. I chose those skippers carefully, only when I was quite confident that they would be 'in' on the joke. The final time I asked about the ganka I foolishly misjudged my informant. He was an owner-skipper who had bought into the industry as an adult, without having apprenticed as a deckie. He was well liked but was different to the others. He did not demand the deference of his deckies that other skippers did; he let them drink some alcohol while laying off the gear, something not done by any other shark boat skipper of which I knew; his number one deckie was a woman, who had taught him most of what he knew about fishing; he ran a charter business to supplement his fishing so was not financially dependent on sharking; he showed little interest in the politics of the industry and he did not smoke, nor drink in the local watering holes as the other skippers did. This should have alerted me to the likelihood of him knowing about the ganka or not. As one who was uninitiated he was not privy to the animal's role in the social and physical environment of the New Jersey waters. But in a moment of – what? Fieldwork fatigue? Stubborn curiosity? – I asked anyway. The result was that the fisherman recognised he had been excluded from this element of the community, which was an unpleasant realisation. Of course, I felt dreadful.

From this point on I began to chastise myself for being so greedy as to want the ganka on display for the whole world to see. I pictured myself as an eighteenth-century white explorer, proudly posing in a yellowing photograph with a rare and wretched animal dangling, lifeless, from the end of my gun. Could my fascination with this animal lead the ganka the way of the dodo? Would searching for this creature destroy it completely?

Conclusion

Gankas climb out of the blackness, across the interface of the living and the dead, and onto shark boats, partly because of a set of social and productive dynamics between shark boat skippers and their deckhands. Like gankas, fishers traverse boundaries that are ordinarily imperforate. In the case of fishers, they corrupt the integrity of the land/sea divide. While most people spend the majority of their productive lives on land, commercial shark fishers are regularly at sea for long periods, isolated from one set of social contexts, but enmeshed in others. In doing so, fishers act as the connection between domains which are, nonetheless, at any given moment, divided. Of course, landscapes and seascapes overlap and are both socialised places. We cannot discuss one without the other, just as the role of the sea creature called the ganka is informed by the productive demands of the community via which it lives. While the immediate function of the ganka in New Jersey waters is to initiate green deckies onto the appropriate rung of the maritime social hierarchy, this creature also acts as a useful metaphor for understanding the ambiguity of those who live their single lives in the very different natural, productive and social environments which emerge for those who are at home on the sea.

On the issue of the ganka as 'trickster' I have two points to make. First, whether or not the ganka can be classified as a 'trickster' is unclear. Based on what little I know about this mischievous sea-creature, it does not appear that the ganka qualifies as a 'trickster' in the way described by Pelton (1980), Teit (1898) or Hynes (1993). This is because the ganka was only ever described to me as being mischievous, and never creative or protective, as archetypical tricksters also tend to be. This brings me to my second, and more important, point. Gathering more data about the ganka might resolve the question of whether or not it is a trickster. Perhaps the ganka does, indeed, display god-like, creative characteristics, along with cheeky, hungry ones. Maybe further inquiry would yield such stories. However, I cannot bring myself to ask New Jersey fishermen about the ganka in case I contribute to it becoming public knowledge within the fishing community. Gankas exist because green deck-hands do not know about them; revealing the ganka would threaten that existence. To delve into the depths of New Jersey waters to discover the breadth and complexity of this creature would be to risk its demise.

Deciding what to do with sensitive or privileged information is not some-thing new to anthropologists; what goes unsaid in an ethnography could fill volumes, and has (Malinowski 1967). In maritime contexts the important information is likely to include specialist knowledge such as carefully guarded fishing spots, and other trade-sensitive knowledge (Blair 2006; Wiber 2002; Wiber et al. 2004). I would argue that knowledge of the ganka is similarly productive information, as it helps shape the behaviour of deckies in a specific productive, social, physical and economic context, and which promotes a more efficient operation, or at least the perception of one. The existence of gankas in New Jersey waters is dependent upon the ecology of those who fish the region. Without the relationships among shark fishermen in which deckies are sometimes made the butt of elaborate jokes, ganka populations would dwindle to the point of extinction. As Douglas (1968: 367) notes in relation to the Norwegian fishermen described by Barth (1966), whose joking relationships emerged only at certain points in the productive cycle; 'the joking ... arises out of the technicalities of fishing'. The ganka is part of a richer world in which shark fishermen dwell, along with boats, fishing gear, mortgages, seafaring novels, wind patterns, bird migrations, family histories, government regulations and buckets of curios. Perhaps, as this world changes, the ganka will no longer be able to survive, or perhaps it will thrive. However, my own role, as one who is not initiated, is to conserve the ganka habitat by keeping the ruse a secret. Obviously (as I have written this chapter), I see academic value in documenting the existence of the ganka. However, to undertake further *research* into the ganka may destroy its efficacy on the wharves, and thus the waves, of New Jersey.

Notes

1 A version of this chapter was first presented at the Association of Social Anthro-pologists annual conference, Bristol University, April 2009. A revised version was workshopped in the Human Ecology seminar series at Rutgers University, December 2010. I am grateful to all those who commented thoughtfully on drafts, including the reviewers and editors of this collection.

2 The exact location, and other identifying details, of the fishing community described in this chapter have been masked considerably in order to protect the ganka.

3 While some zoologically recognised sea-creatures challenge the integrity of the boundary of the sea surface, it is only temporarily. Seals, penguins, and turtles

emerge from the sea for relatively short periods; flying-fish, dolphins and whales breach the surface of the water even more fleetingly. It is precisely these creatures, these 'anomalous water-beings', that often feature in folklore from maritime societies around the world, usually appearing as anthropomorphised protagonists (cf. Bernard, this volume and Schmidt, this volume). For example, the Hans Christian Anderson fairytale, 'The Little Mermaid', which depicts a fish-woman attempting to negotiate both her ocean home and the terrestrial palace of her lover, has been enormously popular worldwide, particularly among Western and Japanese audiences. As with folktales of mermaids and mermen, tales of seal-people or 'selkies' (a term originating in the Orkney Islands) have been recorded widely and reproduced in many societies (see Bernard, this volume and Schmidt, this volume). Selkies are able to temporarily shed their skin and behave as humans on land, donning their skins again to return to the sea.

4 The trickster is often, though not always, male. See Jurich (1998).
5 Indeed, the key boundary that gankas straddle, simultaneously dividing and connecting, is that between the social and the natural environments. As Ingold (1992: 40) suggests, the social and physical are mutually enabling, and we function in a world informed by boundaries of our own fabrication and realisation. These issues have been considered elsewhere (Barth 1969; 2000; Bateson 1972; Cohen 2000; Ingold 2000; King 2005; 2007; Milton 2002; Pálsson 1994; Roepstorff and Bubandt 2003).
6 Such boat-trading periodicals are referred to colloquially by some as 'fishermen's stick-mags', in jocular reference to their insatiable fascination with boats.
7 'Rubber-neck' was one of the nicknames I was given during my fieldwork, a good-natured term meant to capture my position as a 'tourist' in the fishing community.

References

Barth, F. 1966. *Models of Social Organization*. London: Royal Anthropological Institute of Great Britain and Ireland.

Barth, F. 1969. Introduction. In *Ethnic Groups and Boundaries*, edited by F. Barth. Boston: Little, Brown, 9–37.

Barth, F. 2000. Boundaries and Connections. In *Signifying Identities: Anthropological Perspectives on Boundaries and Contested Values*, edited by A.P. Cohen. London: Routledge.

Bateson, G. 1972. *Steps to an Ecology of Mind: Collected Essays in Anthropology, Psychiatry, Evolution and Epistemology*. St Albans: Paladin.

Blair, S.L. 2006. Shooting a Net at 'Gilly's Snag': The Movement of Belonging Among Commercial Fishermen at the Gippsland Lakes. Unpublished Ph.D. thesis. School of Anthropology, Geography and Environmental Studies, University of Melbourne.

Boas, F. 1898. Introduction. In J.A. Teit, *Traditions of the Thompson River Indians of British Columbia*. London: BiblioLife/Houghton Mifflin, 1–18.

Brinton, D.G. 1896. *The Myths of the New World*. 3rd edn. Philadelphia: Sherman & Co. Printers.

Burnham, M. 1998. 'I Lied All the Time': Trickster Discourse and Ethnographic Authority in 'Crashing Thunder', *American Indian Quarterly*, 22(4), 469–484.

Cohen, A.P. (ed.) 2000. *Signifying Identities: Anthropological Perspectives on Boundaries and Contested Values*. London: Routledge.

Davis, W. 2009. *The Wayfinders: The CBC Massey Lectures 2009*. (Accessed 8 February 2010 from www.cbc.ca/ideas/massey/massey2009.html).

Doty, W.G. and Hynes, W.J. 1993. Historical Overview of Theoretical Issues: The Problem of the Trickster. In *Mythical Trickster Figures: Contours, Contexts, and*

Criticisms, edited by W.J. Hynes and W.G. Doty. Tuscaloosa: University of Alabama Press, 13–32.

Douglas, M. 1968. The Social Control of Cognition: Some Factors in Joke Perception, *Man*, 3 (n.s.) (3), 361–376.

Dumont, L. 1980. *Homo Hierarchicus: The Caste System and Its Implications*. Chicago: University of Chicago Press.

Evans-Pritchard, E.E. 1967. *The Zande Trickster*. Oxford: Clarendon Press.

Guenther, M.G. 2002. The Bushman Trickster: Protagonist, Divinity, and Agent of Creativity, *Marvels and Tales*, 16(1), 13–28.

Hyde, L. 1998. *Trickster Makes This World: How Disruptive Imagination Creates Culture*. New York City: North Point Press.

Hynes, W.J. 1993. Mapping the Characteristics of Mythic Tricksters: A Heuristic Guide. In *Mythical Trickster Figures: Contours, Contexts, and Criticisms*, edited by W.J. Hynes and W.G. Doty. Tuscaloosa: University of Alabama Press, 33–45.

Hynes, W. and Doty, G.W. (eds) 1993. *Mythical Trickster Figures: Contours, Contexts, and Criticisms*. Tuscaloosa: University of Alabama Press. Ingold, T. 1992. Culture and the Perception of the Environment. In *Bush Base, Forest Farm: Culture, Environment and Development*, edited by E. Croll and D. Parkin. London: Routledge, 39–56.

Ingold, T. 2000. *The Perception of the Environment: Essays on Livelihood, Dwelling and Skill*. London: Routledge.

Ingold, T. 2001. From the Transmission of Representations to the Education of Attention. In *The Debated Mind: Evolutionary Psychology versus Ethnography*, edited by H. Whitehouse. Kent: Berg, 113–150.

Jung, C.G. 1956. On the Psychology of the Trickster Figure. In *The Trickster: A Study in American Indian Mythology*, edited by P. Radin. New York: Philosophical Library, 195–211.

Jurich, M. 1998. *Scheherazade's Sisters: Trickster Heroines and Their Stories in World Literature*. Westport, CT: Greenwood Press.

King, T.J. 2005. Crisis of Meanings: Divergent Experiences and Perceptions of the Marine Environment in Victoria, Australia. *Australian Journal of Anthropology*, 16(3), 350–365.

King, T.J. 2007. Bad Habits and Prosthetic Performances: Negotiation of Individuality and Embodiment of Social Status in Australian Fishing, *Journal of Anthropological Research*, 63(4), 537–560.

Koepping, K.-P. 1985. Absurdity and Hidden Truth: Cunning Intelligence and Grotesque Body Images as Manifestations of the Trickster, *History of Religions*, 24(3), 191–214.

Lévi-Strauss, C. 1963. *Structural Anthropology*. New York: Basic Books.

Malinowski, B. 1967. *A Diary in the Strict Sense of the Term*. Stanford, CA: Stanford University Press.

McGoodwin, J. 1979. Pelagic Shark Fishing in Rural Mexico: A Context for Co-operative Action, *Ethnology*, 18(4), 325–336.

Milton, K. 2002. *Loving Nature: Towards an Ecology of Emotion*. London: Routledge.

Minnegal, M., King, T.J., Just, R. and Dwyer, P.D. 2003. Deep Identity, Shallow Time: Sustaining a Future in Victorian Fishing Communities, *Australian Journal of Anthropology*, 14(1), 53–71.

Moore, H.L. 1986. *Space, Text, and Gender: An Anthropological Study of the Marakwet of Kenya*. New York: Cambridge University Press.

Ota, Y. and Just, R. 2008. Fleet Sizes, Fishing Effort and the 'Hidden' Factors behind Statistics: An Anthropological Study of Small-Scale Fisheries in UK, *Marine Policy*, 32(3), 301–308.

Pálsson, G. 1991. *Coastal Economies, Cultural Accounts: Human Ecology and Icelandic Discourse*. Manchester: Manchester University Press.

Pálsson, G. 1994. Enskilment at Sea, *Man*, 29(4), 901–927.

Pelton, R.D. 1980. *The Trickster in West Africa: A Study of Mythic Irony and Sacred Delight*. Los Angeles: University of California Press.

Radin, P. 1913. Personal Reminiscences of a Winnebago Indian, *Journal of American Folklore*, 26(102), 293–318.

Radin, P. 1956. *The Trickster: A Study in American Indian Mythology*. New York: Philosophical Library.

Ricketts, M.L. 1966. The North American Indian Trickster, *History of Religions*, 5(2), 327–350.

Roepstorff, A., Bubandt, N. and Kull, K. (eds) 2003. *Imagining Nature: Practices of Cosmology and Identity*. Aarhus, Denmark: Aarhus University Press.

Teit, J.A. 1898. *Traditions of the Thompson River Indians of British Columbia*. London: BiblioLife/Houghton Mifflin.

Turner, V. 1969. *The Ritual Process – Structure and Anti-Structure: A Demonstration of the Use of Ritual and Symbol as a Key to Understanding Social Structure and Processes*. Chicago: Aldine.

Wescott, J. 1962. The Sculpture and Myths of the Eshu-Elegba the Yoruba Trickster: Definition and Interpretation in Yoruba Iconography, *Africa*, 32(4), 336–354.

Wiber, M. 2002. Messy Collaborations: Methodological Issues in Social Science Research For Fisheries Community Based Management, *Max Planck Institute for Social Anthropology, Working Paper no.* 46, 1–32.

Wiber, M., Berkes, F., Charles, A. and Kearney, J. 2004. Participatory Research Supporting Community-Based Fishery Management, *Marine Policy*, 28(6), 459–468.

10 Far from the madding crowd

Big cats on Dartmoor and in Dorset, UK

Adrian Franklin

Introduction

According to recent accounts by social scientists and environmentalists (widely supported by scientists and governmental agencies), most claims and beliefs about the presence of so-called 'alien big cats' (henceforth big cats) in Britain are largely imagined fantasies, social constructions and media-driven hysterias (Buller 2004; 2009; Monbiot 2013). The key elements of this story are as follows: a few big cats, escapees, have been found but a belief arose around the preposterous (and unproven) possibility of their surviving and breeding in the countryside; a number of cult-like cryptozoological groups have formed to study and monitor these animals (developing imaginaries that link these cats to other mythic and primordial bestiaries); the press have picked up on these stories and instead of promoting a healthy scepticism they have reinforced a positive sense of their presence, particularly around the rural hinterlands of cities and towns; and in turn, the constant stream of big-cat stories has served to create a solid sense of their presence in the countryside and thus other animals are mistaken for them in the half-light and shadows, predominantly by urbanites. This desire to believe in their presence draws on very deep-seated meanings and changing relationships between British society and British nature. Such sightings are likened to the long history of imaginary beasts in the UK and a yearned-for return of true wildness. In short, it appears that people now believe them to be there, but more than that, they *want* them to be there, they have become the focus for a new form of *aelurophilia,* or 'love of (in this case feral) big cats'.

According to accounts by Monbiot (2013) and Buller (2004; 2009), in pre-scientific Britain, the countryside resounded with all manner of wild animals, beasts, dragons and dangerous spirits. These gave a sharp definition (an important binary boundary) to a sense of humanity, society and culture, a sense of self and of ontological security. The advent of natural history in the seventeenth century, with its rational and empirical ontology, chased the imagined bestiary away while the extension of hunting and bounties killed the last of the wolves in the seventeenth and eighteenth centuries. With its remaining stock of 'meek' native animals so tightly nestled into the remaining and highly ordered spaces of hedgerow, coppice, woodland and meadow, the UK had become barely wild at all. Then, modern farming methods rendered the countryside even more controlled, tame and domesticated with decreasing amounts of habitat.

> What we are left with is safe and sanitised nature, a complete reversal from the time when the non-urban used to be a wild place, where nymphs lured homebound warriors and where dark forces lay in wait. Pre-industrial civilisation hid behind the city walls. Post-industrial civilisation on the other hand, having tamed nature, spurns the city for the safety of the suburb and country.
>
> (Buller 2009: 10)

However, as cities greened and gained urban forests through suburbanisation and garden city movements in the nineteenth and twentieth centuries, the boundaries between city and countryside, culture and nature began to blur, break down and hybridise. Rare animals like the dormouse (*Muscardinus avellanarius*), peregrine falcon (*Falco peregrinus*) and properly wild animals such as the fox (*Vulpes vulpes*) and badger (*Meles meles*) took refuge in suburban gardens and parks. According to this story, under such circumstances older ontological coordinates of self and society also break down, leading inevitably to the search for order elsewhere. As Henry Buller argues:

> If the human and the non-human elements of the countryside are to be bound together in a new relational hybridity, where should the former look for the necessary boundaries that define their own difference and offer the enduring psychological reassurance of knowing that the 'inside' can only exist if there is an 'outside' beyond it? In other words, *we still need the wild*.
>
> (2009: 13)

Or more specifically:

> these feline messengers reveal our society's essential ambivalence towards the 'Promethean Fear' of Nature's power and ability to resist human appropriation. Nature not only needs to be conceptualised as 'outside' and 'other', it needs to be perceived as such.
>
> (Buller 2009: 23)

So, the vast majority of big cat sightings are not real (though they gain credibility from a few real cases) but an artefact of a deep-seated longing for a natural order to be re-established.

Whatmore and Thorne (1998) and Wolch and Emel (1995) show how recent anxieties about environmental degradation, bio-ethical displacement and species eradications have shifted attention onto defining and legitimating 'true nature' as wildness, and in spatial terms, wilderness (Franklin 2006). This too has driven more vivid binary pairings such as 'them/there – us/here, society-nature, domesticated-wild and so on' (Buller 2004: 14). However plausible, the binaries relevant to anthropological enquiry are always grounded in the culture and society of *specific people*, and the focus and perspective of Buller's (and Monbiot's) analyses are almost entirely focussed on a *generalised* urban population and culture. First, as Buller argues, the alien (i.e. non-endemic) cat phenomenon can be understood by 'modern urban society's impatience with rural nature's all too "human face"' (Buller 2009: 23). Second, he argues that big cat sightings are mostly made by urban people and 'appear to occur most often in the accessible and essentially urbanised countryside of major roads, low-density housing estates and proximal metropolis' (2009: 17). Buller is influenced by the writing of Mike Davis in relation to the mountain lion phenomenon in Los Angeles: 'The Otherness of wild animals is the gestalt which we are constantly refashioning in the image of our own urban confusion and alienation' (Davis 1998: 267 in Buller 2009: 22).

While the evidence shows that indeed some sightings are made in urban hinterlands, it is a gross exaggeration to claim that they are an urban or peri-urban phenomenon. In fact, a substantial number of sightings, possibly the majority, occur in deeper rural locations and most of the hot spots are close to wild areas and away from any major city (e.g. Dorset, the Forest of Dean, Dartmoor, and rural Wales). This means that any analysis which rests on the structural contrast and significance of the urban–wild boundary cannot make an adequate *general* account of big cat sightings in the UK.

There are also problems associated with Buller's view that the big cats 'have meaning' but 'exist [mostly] in non-material terms'. Hurn's (2009) paper reports that there was robust published forensic evidence for the big cats in her locality (Coard 2007; Smith et al. 2006) that substantiated her respondents' view that the big cats are real and known, if few in number (see also Hurn, this volume).

Buller's view is shared by many other commentators, including tellingly perhaps, many governmental and scientific organisations (Wilbert 2006; Hurn 2009). But their 'contestation' should properly be made an element of the analysis, a view to be explored *sociologically*, rather than taken at face value. It is commonly reported that the small number of failed intensive efforts to capture incontrovertible scientific evidence is testament to big cats' mythic status, yet this is never set against comparative scientific research elsewhere. Following up on claims that there are lynx in Dorset, UK, I went to remote Fulufjället National Park in Dalarna, Sweden, a stronghold of the European lynx (*Lynx lynx*). Rangers there told me that they are incredibly hard to see. So hard, in fact, that the regular lynx census exercises have to use reindeer kills as a proxy co-efficient for individual animals. Yet unexplained kills are not conventionally taken as evidence in the UK, although some experts have confirmed big cat predation in this way (see Smith et al. 2006; Coard 2007; Hurn 2009).

In Buller's account attention is focused predominantly on human sightings and the dialogue and myth making that stemmed from them. The fact that there are instances of recorded and scientifically validated cases of non-native cats at large in Britain is not adequately dealt with, and particularly so in instances where the scientific/governmental communities (who have largely supported the 'myth' hypothesis) have been shown to hide firm evidence from the public gaze. The fact that very significant numbers of non-native cats were imported and kept in private collections, zoos and as individual 'pets' remains at once acknowledged and detached from the 'mythic sightings' narrative.

Buller recognises that the myth-making is based on extrapolations from a few plausible or proven cases of big cat escapees (and here he makes much of Serres' interest in how the real and imaginary take on an independent life of their own as intermediaries or 'angels', to use Serres' [2003] term), but throughout his article he is predominantly scathing of claims of their more widespread presence. Nonetheless he is not above exaggeration himself – at one point, for example, he notes that 'it is rather comforting to learn that the English countryside offers "optimum conditions" for the reproduction of lion and tigers', but this was clearly included to discredit big cat sightings and organisations rather than reflect fairly on their claims which barely ever mention these species. Curiously, biologist Ross Barnett used the same exaggeration to discredit significance being attached to the story of a puma (*Puma concolor*) that was captured in Inverness-shire in 1980. He said: 'It's all very good saying you saw a lion in Essex or a tiger in Shropshire, or wherever. But it is very difficult to estimate size of a species from a distance – especially if you are unfamiliar with them' (Morelle 2013). Why refer to animals that are barely mentioned in big cat sightings, if not to cast doubt on the veracity of reporting?

Second, Buller's socio-cultural environmental history is sketchy at best and there is a failure to explain why it is that non-native and very exotic cats could be deemed *fitting* animals to restore a sense of lost wildness. If such a longing existed, then the British would be surely be doing what the Swedes, Canadians and Norwegians have done: to restore their own 'dangerous wild animals' in the form of wolves (*Canis lupus lupus*), boars (*Sus scrofa*) and bears (*Ursus arctos arctos*) (Marvin 2011a; 2011b; Pretty 2007; Whatmore and Thorn 1998). However such enthusiasm is, at best, half-hearted (Marvin 2011a).

Ritvo (1987) has shown how Britain's long legacy as an imperial culture has had a crucial bearing on its human–animal relations. The apparent British liking for big cats may be explained in the same way they have embraced and tolerated other exotic species in contemporary landscapes (e.g. muntjac deer [*Muntiacus reevesi*], parakeets [*Myiopsitta monachus* and *Psittacula krameri*] and wallabies [*Macropus rufogriseus*]). As an imperial culture predicated as much on acquiring and adding new entities to its sense of self as it was on defending its indigenous order, the popularity of importing big cats, zoological gardens and tolerance of established wild populations is perfectly consistent.

British imperial culture was, in many senses, a recovery of classical models that sought the *civilisation* of wildness rather than its glorification, and this was born out by British enthusiasm for the acclimatisation movement in the nineteenth century. In Australia for example, there was an audacious reordering of nature where (choice) imported species were cultivated and favoured in place of (inferior) indigenous species (Franklin 2012; Ritvo 2010). Since the Romano-British period, British nature was steadily added to – a process that accelerated considerably after the sixteenth century (Thomas 1983) – without much in the way of reaction or reversal.

Another possibility is that this alleged *desire* for big cats is misplaced or simply wrong. Perhaps, they really *are* there and a belief is sustained through regular sightings of them. This is clearly the case for the large cluster of sightings recorded for the remote Forest of Dean in Gloucestershire during the 2000s. These also remained speculative and 'mythic' until, under a Freedom of Information request to the Forestry Commission (FC), it was 'revealed by the scientists that the government agency confirmed that two "reliable" sightings of large cats have taken place in the last seven years' (Woodward 2009). Why were they not reported in the proper way? Could it possibly be because the Forest of Dean is a major commodified leisure facility and an important investment of the FC? They know that people are actually sensible: they don't want big cats in the forests where their children come to learn and play.

The 'longing for the wild' thesis is problematic because it rests too much on reported sightings and newspaper accounts *and* places too much emphasis on their urban origins. In Hurn's (2009) case a different story emerged because instead of using sightings as primary data, she embarked on a qualitative investigation *in the area where they were seen*. In this case, the sightings could be related to beliefs and knowledge that only made sense once they were placed in their wider social and historical context (see also Hurn, this volume). It begged the question as to whether this was a one-off or whether similar results might be found if a similar investigation were conducted elsewhere.

Findings from a locality-based approach in Dartmoor, Devon and North Dorset

Fieldwork was conducted in two case study areas in a rural (North Dorset) and remote (Dartmoor) area of England for ten weeks in spring 2011. An ethnographic approach was adopted in which I visited particular villages and hamlets (where sightings had been reported) with the aim of talking to as many people as possible, rather than conducting formal interviews. In the very gregarious spaces of pubs, cafes, footpaths and local stores it is easy to fall into conversation with people and gauge how significant big cat sightings have been and what local people think about them. Forty significant conversations were recorded and these were augmented by a wide range of other exchanges, observations and more formal talks with local experts and professionals. Walking the local footpaths also gave an understanding of the environment in which sightings take place and assisted in following conversations set in local geographies. Tourism research (e.g. Ryan 1995) has shown how the researcher can participate in a locality as a visitor and still produce useful observational and conversational data while not actually staying for significantly long periods or conducting formal interviews, especially where interviews might be difficult to arrange spontaneously among people moving through landscapes. Data from a similar conversational methodology usefully informed Hurn's study of big cat encounters in Wales (Hurn 2009: 9), although Hurn was a long-term resident and immersed in the life of the local community.

I visited a few villages and small towns on the upper slopes and high country of Dartmoor over a period spanning January to mid-February 2011, and in North Dorset over two periods in April and June 2011. In both localities tourism has been a valued and essential feature of the local economy for over 100 years and so an enquiring tourist is neither unusual nor avoided. The ubiquitous presence of people with either dogs or in clothes that indicated they had been working with farm animals made beginning conversations about big cats very easy. I could open conversations by expressing concern about the dangers of big cats to their dog or lambs.

Dartmoor

Dartmoor is an upland area of some 950 square kilometres in south Devon, England, which is protected by its National Park status. It is an ancient weathered mass of granite from the Carboniferous period forming a very wild and exposed upper expanse of moor reaching to 621 metres above sea level, with richly wooded lower slopes. It is sparsely populated across its high saddle but has a series of thriving towns around its lower points, most of them medieval stannary towns. Licensed by the Crown to produce and distribute tin, they became wealthy centres of tin mining, sale and minting with their own parliaments.

In Dartmoor it was striking how most reporting hot spots were not out on the open country of the moor (which has a lot of visitation from city-based tourists) but closer to areas less popular with tourists where the moor meets denser woodland and more mixed habitat/cover. At first, there also seemed no reason why there should be more sightings in the northeastern sector that included the towns of Chagford, Moretonhampstead, Postbridge and Okehampton. But when one walks there and looks closely at Ordnance Survey maps it is apparent that this area is adjacent to the greatest stands of forest, particularly at Fernworthy (pine) and west of Chagford (very mixed). Feasibly, big cats could be well hidden and adequately nourished in such landscapes.

Fieldwork was carried out on the lower-slope towns of Moretonhamstead, Okehampton, Chagford and the higher village of Postbridge. These were selected for their consistent records of big cat sightings in the areas around them, not least in reports to police that emerged from Freedom of Information enquiries. Conversations confirmed the notion that sightings were more concentrated than would be the case if they were merely the result of generalised and wilful imagination. For example, at Okehampton, a small town with lower lying and more open farmed country around it, conversations in cafes, charity shops, and on the street produced no sightings stories. At the local library a librarian also had little to report. At a veterinary practice, vets confirmed no reports of local sightings or attacks on livestock or companion animals that were attributed to big cats. However, they did tell me there *were* many such reports further west, in Chagford and also at Castle Drogo.

At Chagford, 9 February 2011, 11 a.m.

Chagford is an exceedingly wealthy Devonshire town. It has many famous celebrity residents, including British comedian Jennifer Saunders and her husband, the comedian Adrian Edmondson. Its buildings are in an opulent medieval style and not what you would expect in such a remote area. Most of the official and municipal buildings, including the parliament building, are set around a central square. There is one pub on the square that I entered at mid-morning for a coffee. I noticed a group of seven senior citizens also having coffee. By their accents and dress I could infer they were Devonshire rural middle-class people, possibly landowners or farmers. One couple had a dog which I made a fuss of and we fell into conversation for a while as the group sat down at a nearby table. I mentioned talking to a villager with a small dog who had sighted a big cat and this created a lively conversation with a wide variety of stories and rumours recounted. But one man remained sceptical and laughed at the very notion, and his assertive scorn soon closed the conversation down. However, as I remained standing by the bar, one by one the other members of the group came forward to tell me their big cat story, reluctant, it seemed, to do so more publicly. One retired woman had a small white dog and had been used to taking it for a walk, beginning at the village church wicket gate that opened onto a wooded hillside. She claimed to have seen a big cat there and no longer ventured outside the built area of the village.

A more elderly woman in her eighties who had remained quiet during the conversation around their table came up to tell me that she had been brought up in the local manor and that as a girl she was frequently taken to London and shown the exotic animals on sale in Harrods. She said these included many big cats and that in the 1930s most of the big houses in the locality had small menageries that included big cats, including her own house. Then she told me how, during the Second World War, when only the women and a skeleton staff of other women were left, they decided to liberate the big cats because there was not enough food to feed them. She said that she has felt terrible ever since, at first for the welfare of the cats, and later for people and farmers when the first sightings of 'wild' big cats occurred. To obtain such a 'foundation' story after hearing so many sightings stories was not anticipated but it highlights something that is not at all unusual or remarkable. The wealthy landowning class was both numerous and evenly scattered, particularly across southern England, and they had a well-known liking for keeping exotic animal menageries, no doubt emulating the major royal and courtly homes. There was an established and extensive trade catering to a nation-wide obsession for private menageries: Harrods, for example, opened an exotic pets department in the 1890s. As early as the 1760s

> a distinct geography of animal exhibitions and commerce had emerged in London as animal merchants and menageries lined the Strand, Piccadilly and St James's. In the following decades and certainly by the 1800s they would proliferate to the extent that it was possible to walk through West London and see all the principal 'animals of importance'.
>
> (Plumb 2010: 55)

This barely regulated trade continued until 1976 when an Act of Parliament made the keeping of such animals prohibitively expensive. At this time, it is alleged that a lot of owners liberated them rather than upgrade their accommodation and security to that required by the new Act (e.g. Hurn 2009). If the extent of big cat sales could be verified, and if menageries as fashionable cultural or status symbols could be linked to specific country houses in certain districts, it is not only possible but very likely that local people experienced the presence of escapees and liberations over some considerable period of time. For example, in 1903 a Canadian lynx (*Lynx canadensis*) was donated to Bristol Museum and Art Gallery, where it has been in store ever since. According to Morelle (2013) 'The museum's records state that it had been shot after attacking and killing two dogs close to Newton Abbot in Devon.' Newton Abbot is a mere 19 miles or a 35-minute drive from Chagford on the edge of Dartmoor, and certainly within the district referred to by the respondent above. It would not take very many such animals to create sufficient sightings for them to become an entrenched element of local wildlife, and would not require the animals to establish a breeding colony.

Later that evening in another pub, the young woman serving behind the bar asked me why I was visiting Chagford and I told her about my interest in big cat sightings. She told me that this was a major topic of conversation and that she and her girlfriend had seen a big cat at close quarters while walking back from a village pub nearby. She reported being so scared that they had to call her father to come and pick them up. This was about a year before. The story seemed genuine, but its veracity is less important than how it relates to a body of local stories and local knowledge about an environment in which there were big cats, or there had been and *could* be still. This is very different to a media-driven imaginary among urbanites who become excited at the prospect of true wildness returning to the outer areas of their everyday. For this woman, the encounter *was* her everyday and it was not romantic but scary and unnerving. She may have taken pleasure in relating the story but I did not detect the tell-tale sign of sensation, rather the more practical tones of risk management.

At Postbridge Post Office, 7 February 2011, 10.30 a.m.

Postbridge is a small village that lies between the lower and higher slopes of Dartmoor, and with its thirteenth-century clapper bridge built to carry tin across the East Dart River is a popular stopping point for tourists. It is in an area where several big cat sightings had been reported. I entered the Post Office and had a look around its general stores facilities, its cream tearooms beyond, and through its postcard stand. Were there cryptozoological cards of the big cats of Dartmoor for tourists to buy, maybe? No, there were not. Despite the number of sightings and the growing fame of Devon's moorland 'beasts', it had not been commercialised at all, certainly not along the lines of the Loch Ness Monster, for instance.

There were two other people in the Post Office. A woman in her fifties was perusing cards and another a woman in her sixties was transacting business at the counter. Both women were in working attire. The former was dressed in horse riding gear while the latter was dressed in warm clothing with clear signs of having been handling sheep. As the woman in shepherding gear completed her transaction I was behind her waiting to pay for some postcards. She had a small spaniel with her and I said, 'I bet you have to be careful with her – what with all these big cats around here!' It was a conversation starter I invariably used (everyone had a dog around here) and was designed to suggest the inevitable curiosity of visitors to this area. She startled me by saying 'Don't you joke about it young man, we've had them on our farm and it is not a laughing matter!' She went on to recount how one day her mother had called her to see a huge cat that was just outside their kitchen window near some penned sheep. She raced to open the door and the cat jumped up onto a wall and ran out behind the house and into cover. She had seen it from no less than five metres away. It was sandy coloured and a very large animal, the size of a large dog. 'But a dog could not have jumped onto that wall, moved like that or looked like that.' She had not seen it again.

By this time the other woman, who was not known to the first, had come to pay for her card. She had heard the conversation and she joined in:

> We have had them over at our place, for sure, they have been seen and actually one of our ponies was attacked and seriously injured. The vet could not understand the injuries and nor could we. Not a dog, definitely not a fox. There is not a lot it could have been besides these!

Here were two seemingly credible stories from a visit to one very remote Post Office in the middle of nowhere; both appeared to be sensible, educated women – typical small farmers in this area. However, neither incident had been reported.

Then the postmistress began her own story. She and her husband had a farm nearby and they saw a big cat at close range near their new chicken shed and it had actually left a paw mark on some setting concrete. This was not reported either. She then recounted several other stories that had been told to her over the counter. The Post Office being something of a local epicentre, she had collected incidents from places in all directions, and Chagford was mentioned several times. I tried to summarise what the women had said to me. Did they think that big cats were widely experienced in the locality, that this morning's meeting was not just a lucky coincidence for me? 'Yes', the postmistress said, 'you will find that more or less everyone around here has seen or heard of something'. I left feeling that I should return and spend more time following up these accounts, and paw prints!

Dorset

The county of Dorset is the easternmost part of the 'West Country' (i.e southwest England) and ostensibly has changed very little in the twentieth century. It is one of the few counties not to have a major motorway running through it, and most of Dorset grades down to the roads created in the early part of the twentieth and the latter part of the nineteenth centuries. In agricultural terms it is a dairy and mixed farming area with a great deal of pasture set on its open chalk downland. In addition to very extensive woodland there are also huge swathes of Forestry Commission pine forests. Dorset was the only county to have its own active big cat organisation (Dorset Big Cats) and its detailed information about sightings was a starting point for choosing a locality to study. Three areas stood out for their intensity of sightings. The first is the Stour valley from the north, between Sherbourne and Shaftesbury, to the southern villages around Blandford Forum. The second is the River Frome valley that describes a line passing through Dorchester on a northeast–southwest axis. The third is the countryside around Bridport.

These sightings are very clustered (see Harpur 2008 for a map), far from spatially random sightings, and the territories they describe are very specific. The country around Bridport is very sparsely populated in every direction, it has a wide diversity of habitats and it has a great abundance of food sources, both wild and domestic. Walking around in this part of Dorset one is struck by the absence of humanity for most of the day wherever one walks. The Frome valley is also extremely diverse, with an abundance of cover, year round water and corridors of country linking areas of woodland. The Stour valley has a similar network of corridors between forests and woods but it has much larger tracts of woodland, and also wilder uplands on its line of chalk hills, most of them former Iron Age hill forts. Walking around this area I saw an abundance of rabbits (*Oryctolagus cuniculus*), badgers (*Meles meles*), roe deer (*Capreolus capreolus*) and foxes (*Vulpes vulpes*) as well as domestic sheep and cattle.

I decided to home in on the Stour valley because of a spate of sightings that had occurred in recent years, in addition to the consistency of these sightings. Almost invariably the cats seen were large and dark. The other intriguing thing is that almost all of the sightings had been made by locals such as farmers and other native Dorset people as opposed to the group who are slowly displacing them as residents: affluent middle-class Londoners seeking either a holiday home or a move to the country. Some idea of the extent of this in-migration was given by a village doctor who had lived in one of my case study villages for most of his professional life. In the parish magazine he wrote an article about how the village had been transformed from a working village with its manor house, its farmers, its small professional group and its majority of skilled and unskilled rural workers and small businesses, to a village of largely part-time middle-class residents, urban retirees from London and a group of working middle-class professionals. Only the farming families had remained on the land and it is mostly they who are still working there in the early morning and late in the day when most sightings are made.

In this particular village one of the farmers walked around his land twice a day, every day. He knew every inch of that landscape. He was born there and had only ever worked there. Knowing the local wildlife intimately he was also a big cat sceptic until one day, he saw a large black cat running across the yard at the back of his house. It was not a great distance and he saw it very well. He did not report it and felt reluctant to talk about it, fearing that others would not believe him. That is, until he spoke to a friend, a local retired electrician, who said that on the same day he himself had seen a large dark cat some thirty metres away. His friend had the advantage of seeing it against the grid of a wide, five-bar field gate and he reckoned it was about the size of a large dog with a very long tail, thicker at the base than that of a domestic cat. Like the farmer's sighting, the cat was soon away into the tall grasses and hedges that lined the path. From there, this creature, whatever it was, was a mere 100 metres away from the cover of a copse, and beyond that a corridor of woodland that stretched deep into uninhabited hills. On two subsequent walks around this area I made two observations. The first occurred on a very early morning walk out of the village and up high into the chalk hills above it. When I reached the summit, I saw a fox in the distance, a large male in peak condition. Although the light was far from good I was able to make this accurate assessment from at least 400 metres away, possibly more. It made me realise that the farmer and the electrician who had been looking out onto this hill all their lives would surely not mistake a big cat for some other animal. The second observation was that this entire area had interlocking woodland, some of it particularly dense, overgrown and impenetrable; certainly a habitat that could conceal a big cat.

The electrician had never reported his sighting. Having got to know these two men, their reputations and characters, it was very hard not to believe their stories. In both cases they were natural sceptics, plain talking and not prone to exaggeration: in fact the opposite was the case, they were men of few words but what they said was intelligent and insightful. Neither were aware of another story involving another black cat and a farmer on one of the nearby chalk hills, no more than five miles away. A roe deer carcass was found some eight feet into the fork of a large beech tree in one of this farmer's woods, where it subsequently rotted. This caching of prey with the intention of removing it from other predators would indeed be the style of a medium-to-large felid. According to the farmer, he permitted deer stalkers to keep roe deer numbers down, and a stalker he spoke to said that they do occasionally leave a carcass up a in tree until they have time to retrieve it. However, without a ladder no hunter could have lifted a deer so high, begging the question, how did it get there? He did not have an internet connection and was unaware of seven other sightings in the immediate area as reported only to Dorset Big Cats.

In this district I spoke to only one person who had no direct or indirect experience of big cats. Everyone else (fourteen individuals) had a story. Nobody I spoke to had reported their sightings to the police or the newspaper or to Dorset Big Cats and, like the farmer and electrician cited above, most people were somewhat reluctant to talk about their experiences openly. I heard that many farmers keep sightings quiet because they wished to be able to kill the animals without interference from protective authorities (cf. Hurn 2009).

When I asked a Dorset Wildlife Trust officer working in a nearby town whether he believed there were big cats in the area he was unequivocal.

> They are real alright. I know a lot of the people who have seen them and they would know very well. They are country people; they know all the native wildlife and would soon spot something unusual. I know their individual powers of observation and knowledge so they are trustworthy. Plus they have no reason to lie [...] They are European lynx mainly, a few pumas and regular sightings of black panthers. I think a lot of them are coming over from the New Forest, coming into Dorset via the area around Ringwood. They are up the Stour valley and especially in and around those huge Forestry Commission forests in the Wareham area [...] There's about eight solid sightings a year, and most of those have photos [...] they are consistent with big cats but they never get close enough for an ideal photo; well, these cats are so elusive, so scared of humans they always see or sense you before you see them and so inevitably the photos are almost always from some distance. But you can tell even with these by the thickness of the tail as a proportion of their body. They could not be domestic cats or dogs, you can see that. Also the tails are too long and carried in a different way or short in the case of the lynx. Most of them are pale or lightish coloured, so that's consistent with a European lynx or puma. Many appear black [...] there have been bodies found, often they are unhealthy with mange and very thin; some of them there's nothing to them [...] they're probably too ill to breed. Some have been found by Forestry Commission photo census of deer; most have been taken by the big cat societies; if they get wind of a sighting they will get in there and look for evidence of their tracks or kills and they will set up trip-wired

cameras all over [...] they want to know as much as they can you see, document them here.

There is also a hunting and fishing store locally and the workers there are a good source of information on local wildlife. Have hunters reported seeing big cats in the area? The joint owners contributed to this summarised response:

Oh yes. A lot of our clients have seen something over the years but very recently a really good hunter, one of our regulars, was actually attacked by one. Quite seriously actually, had to be treated in hospital. He was walking back from a rough shoot and he had some game on him and he reckons he was stalked. Anyway, the injuries were quite ferocious. This was over Wareham way, in among those big Forestry Commission forests. He had seen cats in there before but this was his first close encounter sort of thing! He is very shaken up. Could not have been a dog because of the claw marks and tears and it was too big for a domestic cat. He reckoned it was a lynx.

The truth of these sightings is arguably less interesting here than their meaning for local people. The respondents I spoke to in Dorset gave no indication that these animals represented a yearned-for return of true wildness. Generalising across their responses one would have to conclude that the most common and immediate response was that these animals could mean trouble. This was clearly evident in the tone of the men in the hunting shop and it was implicit in the accounts of others. There are serious consequences to having a large predator on the loose, presenting a hazard to human as well as non-human life. Therefore, perhaps unsurprisingly, the overwhelming response to the presence of big cats was negative, albeit accompanied by the belief that there cannot be many of them, and the knowledge that they are secretive and mostly nocturnal.

What respondents think about these sightings arguably depends on how they explain their origins. A breeding population embedded as a feature of the countryside is an altogether different problem to the transient presence of a few rogue individuals who have escaped from a zoo or some other source of captivity. The people I spoke to ranged from those who had not really thought much about the source of big cats in the area to those who subscribed to the latter view. It is common knowledge that many rock stars live in Dorset and Gloucestershire, and I also encountered unverified rumours that they and some remote-living drug dealers have kept big cats at various times. This area of Dorset also has a very large number of manors and, as noted earlier in relation to Chagford, many of these would have had menageries in the past. However, there is a widespread sense that this is a recent, post-1970s phenomenon in the wake of the Dangerous Wild Animals Act which came into force in 1976. So big cat sightings could be consistent with changes to the legal status of these animals. There is no doubt that people enjoy telling (and hearing) sightings stories but one is also aware that despite the reality of sightings for those who experience them, they also exist in a general condition of *disbelief*. The sighters and those receiving the story are never quite fully convinced, even though they cannot find any other way of describing what they saw.

Conclusion

This chapter challenged the hypothesis presented in existing academic accounts (e.g. Buller 2009) that the feral big cat phenomenon in the UK is largely an urban fantasy stimulated by a crisis in ontological security. According to such a view, an external wild nature that had always provided shape and definition to civil sensibilities and society has been steadily eroded, leaving people confused and disorientated. The recovery of wildness through the continuing presence of potentially dangerous big cats offered the possibility of redemption, a return to the true order of the world where humanity was once again checked and balanced by nature. This thesis draws on the very widespread enthusiasm for wildness, wilderness and the re-wilding of environments in modern nations around the world, as well as on the notion of deep historic pasts, inhabited by monsters and beasts. It offers a feasible account of big cat sightings as mythic rather than true and as belonging to urban rather than rural cultures.

I have argued that this thesis has many flaws. Taking evidence mostly from town- and city-based local newspaper reports (Buller 2004; 2009; Monbiot 2013) results in an urban bias, which, as has been shown, is not reflected in the true pattern of sightings, reported or otherwise. In fact, this pattern shows very intense concentrations in remote and rural parts of England that would not fit the urban thesis or a more generalised pan-British 'fantasy' thesis (if the same sentiments might be claimed for the entire population). Further, such an argument fails to consider why the UK has been largely unenthusiastic towards re-wilding policies involving its own indigenous large carnivores such as wolves, bears or lynx, thus casting some considerable doubt as to why non-indigenous animals from South America, Africa, or North America might be deemed more relevant or acceptable.

There are other possible explanations that better fit the evidence, particularly the historical precedent for non-endemic big cat ownership in rural areas. Indeed, the re-wilding thesis overlooks Britain's historical status as an *imperial* culture and in particular, the tendency of the landed gentry to add or collect 'exotic' cultural and natural elements as part of their social identity. The thesis also downplays or ignores the narratives surrounding how they potentially became embedded in the English countryside, initially as parts of menageries and then as escapees or releases. There is no doubt that a trickle of escapees would have been a feature of such private collections, as would releases during times of hardship and in the 1970s following new legislation related to keeping such animals in captivity (see also Hurn 2009).

The chapter reported on fieldwork conducted in two very different rural localities where there was elevated and spatially concentrated reporting of big cat sightings (in addition to a large number of unreported sightings which only became apparent during the course of empirical research). These places were characterised by their proximity to environments suitable for medium-to-large non-endemic cats such as lynx or leopards and a local history of their presence there. Most local people encountered in these places had direct experience or sightings of big cats themselves or had encountered these animals indirectly through family and friendship networks. Their presence was mostly regarded as a reality and not particularly remarkable, to the point where many encounters were not reported to the authorities, but they certainly were not regarded with any great enthusiasm. Predictably, for cultures and individuals with instrumental and less romantic views of the countryside, practical matters weighed most heavily when it came to attributing meaning to sightings, with the balance of opinion erring on the side of inconvenience and risk. Since livestock were believed to be attacked, companion animals were thought to be at risk, and incomes made from tourism might also be threatened by the presence of large predators, reporting to authorities, active big cat enthusiasts and the media made little sense. Reading between the lines, I felt these individuals wanted to manage big cats, if they ever came across them, in their own way. Hence, subsequent research might find rural under-reporting in addition to urban over-reporting. However, meanings and actions are always likely to be locally contingent with no overarching national response

In this volume and in a previous publication Hurn (2009) reports evidence of sightings from Ceredigion, Wales. In remote West Wales 'sightings seemed to be rather common' and her case study suggests an entirely different reason why rural people may view big cats positively. In this context the positive identities ascribed to big cats derived from their identification as a symbol of resistance and liberation for the local Welsh community against historical and contemporary interference from outsiders (especially the English) and a natural antidote to an unwanted animal that was associated with those human outsiders: the fox.

Hurn suggests that the meanings of big cat experiences are socially contingent and symptomatic of the wider historical and social milieu of the localities into which big cats enter (or are seen). Her analysis reminds us that in matters of 'nature' and 'environment' meanings are rarely shared in the manner suggested by Buller's grand (and essentialist) narratives of urban/domestic versus rural/wild imaginaries. Rather, as McNaughton and Urry (1998) observe, they typically hinge on locally specific, contested natures and social conflict.

More work remains to be done but the results of this preliminary fieldwork undermine the commonly held 'mythical' view of the big cat phenomenon in the UK. They do not prove the existence of big cats so much as cast doubt on their purely imagined existence. There may be a very small number of them but it would only take a very small number to create an enormous *cultural* presence. Following Hurn (2009) this chapter suggests that views about big cat sightings are likely to depend very much on the *contingent* way they configure within the historical and contemporary forms of human dwelling in the countryside (see Ingold 2000). The interview data and sheer volume of sightings supports Hurn's assertion that big cats have to be taken more seriously, and highlights the dangers of falling back on basic and simplistic binary oppositions when attempting to explain the complexity of human relationships to, with and in natural environments. Most of all it reaffirms the value of qualitative fieldwork and ethnography as a corrective to armchair theorising.

References

Buller, H. 2004. Where the wild things are: the evolving iconography of rural fauna. *Journal of Rural Studies*, 20, 131–141.

Buller, H. 2009. The Woozle's return: angels, monsters and imaginary spaces. Geography Department workshop paper, University of Exeter.

Coard, R. 2007. Ascertaining an agent: using tooth pit data to determine the carnivores responsible for predation in cases of suspected big cat kills in an upland area of Britain. *Journal of Archaeological Science*, 34(10), 1677–1684.

Franklin, A.S. 2006. The [in]humanity of the wilderness photo. *Australian Humanities Review*, 28 April, 1–16.

Franklin, A.S. 2012. An Improper Nature? 'Species Cleansing' in Australia. In *Human and Other Animals: Critical Perspectives*, edited by B. Carter and N. Charles. London: Palgrave Macmillan, 195–216.

Harpur, M. 2008. *Roaring Dorset: Encounters with Big Cats*. Dorchester, Dorset: Roving Press.

Hurn, S. 2009. Here be dragons? No, big cats! Predator symbolism in rural West Wales. *Anthropology Today*, 25(1), 6–11.

Ingold, T. 2000. *The Perception of the Environment: Essays in Livelihood, Dwelling and Skill*. London: Routledge.

Macnaghten, P. and Urry, J. 1998. *Contested Natures*. London: Sage.

Marvin, G. 2011a. Wolves: now at our door. *Guardian*, 27 July. (Accessed 3 November 2015 from www.guardian.co.uk/commentisfree/2011/jul/27/wolf-at-door-home-counties).

Marvin, G. 2011b. *Wolf*. London: Reaktion Press.

Monbiot, G. 2013. *Feral: Searching for Enchantment on the Frontiers of Rewilding*. Harmondsworth: Penguin.

Morelle, R. 2013. 'Big cat' Canadian lynx was on the loose in UK in 1903. *BBC TV News*, 25 April. (Accessed 25 June 2013 from www.bbc.co.uk/news/science-environment-22263874).

Plumb, C. 2010. Exotic animals in eighteenth century Britain. Unpublished Ph.D. thesis, Centre for Museology, University of Manchester.

Pretty, J. 2007. *The Earth Only Endures*. London: Earthscan.

Ritvo, H. 1987. *The Animal Estate: The English and Other Creatures in the Victorian Age*. Cambridge, MA: Harvard University Press.

Ritvo, H. 2010. *Noble Cows and Hybrid Zebras: Essays on Animals and History*. Charlottesville: University of Virginia Press.

Ryan, C. 1995. Learning about tourists from conversations. *Tourism Management*, 16(3), 207–215.

Serres, M. 2003. *L'incadescent*. Paris: Editions le Pommier.

Smith, A.B., Street-Perrott, F.A. and Hooper, T. 2006. A Method for Grading Sightings of Non-native Cats: Application to South and West Wales. In *Proceedings of the Eastern Cougar Conference 2004*, edited by J. Tischendorf, H. McGinnis and S.J. Ropski. North Spring, WV: Eastern Cougar Foundation and the American *Ecol Res* Institute, 102–121.

Thomas, K. 1983. *Man and the Natural World: Changing Attitudes in England 1500–1800*. London: Allen Lane.

Whatmore, S. and Thorne, L. 1998. Wild(er)ness: reconfiguring the geographies of wildlife. *Transactions of the Institute of British Geographers*, 23(4), 435–454.

Wilbert, C. 2006. What Is Doing the Killing? Animal Attacks, Man-eaters, and Shifting Boundaries and Flows of Human–Animal Relations. In *Killing Animals*, edited by the Animal Studies Group. Urbana: University of Illinois Press, 30–49.

Wolch, J. and Emel, J. 1995. Bringing the animals back in. *Environment and Planning D: Society and Space*, 13, 631–730.

Woodward, J. 2009. 'Reliable' big cat sightings revealed. *Independent*, 7 January. (Accessed 25 June 2016 from www.independent.co.uk/environment/nature/reliable-big-cat-sightings-revealed-1229204.html).

11 Land of beasts and dragons

Contemporary myth-making in rural Wales

Samantha Hurn

Myth as it exists in a savage community, that is, in its living primitive form, is not merely a story told but a reality lived. It is not of the nature of fiction, such as we read today in a novel, but it is a living reality, believed to have once happened in primeval times, and continuing ever since to influence the world and human destinies. This myth is to the savage what, to a fully believing Christian, is the Biblical story of Creation, of the Fall, of the Redemption by Christ's Sacrifice on the Cross. As our sacred story lives in our ritual, in our morality, as it governs our faith and controls our conduct, even so does his myth for the savage.

(Malinowski 2004: 78)

Ever since Bacon, science has insisted on discovering the truth of what is there, and thus on the strict separation of fact and interpretation.

(Ingold 2013a: 745)

Introduction

There are several themes to be addressed in this chapter. First, liminality was a key issue which surfaced repeatedly during the course of my research into so-called Alien Big Cat (ABC) sightings in Wales, UK (where 'Alien' reflects the animals' non-endemic status). The unknown nature, origin and behaviours of the cats; the presence of these creatures in a landscape which is not their native 'home'; and the difficulties associated with verifying their presence are all factors which contribute to the status of ABCs as 'matter out of place'. The question of authority in relation to the validity of sightings has also come to the fore more recently, and this is another theme which will be considered here. I argue that authority (in the sense of whose accounts are credible) is linked to the liminal or transgressive nature of ABCs, (and this applies also to cryptozoological entities or 'cryptids' more generally). Their unknown and unknowable status makes them risky subjects for authority figures to deal with and a balance is sought between the reality of a sighting as a visceral, tangible experience for the individual in question, and the need to account for what could otherwise be dismissed as the product of an over-active imagination. In his recent writing, Ingold (2013a and 2013b) has argued that engaging with cryptids (although he doesn't refer to them as such) can heal what he sees as the rupture 'between the real world and our imagination of it' (2013b: 749). By taking the existence of ABCs seriously I seek to reflect on the uneasy relationships between anthropology and relativism on the one hand, and science and imagination on the other, both of which are key aspects of the recent ontological turn in anthropology.

The symbolism of ABCs

ABC sightings in the UK, as with sightings of or encounters with most cryptozoological entities, are the stuff of myth, legend and contested narrative. Malinowski, one of the founding fathers of British social anthropology, recognised the 'function' of myths in all human societies; myths are useful vehicles for encoding and expressing information which is important within any given socio-cultural context. Contemporary myths, such as those which have arisen around cryptids, can also be seen to serve important social functions. Sometimes cautionary or moral tales, they can also expose the relationships between different strata of the society, provided, as Douglas (1996) cautions, they are considered within their specific social and historical context (see also Attala; Bernard; Heneise; High; King; and Merz, this volume).

There is precedent for cryptids playing a symbolic role in the complex relationships between Wales and England in the UK, between an 'internal colony' (Hechter 1977) and the sovereign power whose laws impact on the lives and livelihoods of the colonised. While the patron saint of England, St George, is credited with slaying a dragon, the iconic Welsh dragon of the national flag represents the power of the tribes of Wales in the past and the pride and resilience of the country's contemporary inhabitants. Dragons, or their 'known' palaeontological equivalents[1], no longer roam the Welsh valleys and hillsides, yet something of *y draig*'s enigmatic ability to mobilise and fortify the nation is retained, or rather reinvented, in contemporary myths surrounding ABC sightings. When viewed in such a light, the 'reality' of big cats is arguably less important than their role as trope or archetypal figure in a politically charged narrative concerning the place of marginalised rural communities in a globalised world. Indeed, contemporary myths are often reflections of real life concerns as exemplified in Scheper-Hughes' (2001) research on 'body snatchers' and the illegal trade in human organs, and West's (2005) discussions of the were-lions who terrorised politically disenfranchised rural peasants in post-war Mozambique. Consequently, characters such as ABCs might be regarded as manifestations of the fears associated with change, resistance or a loss of control (or all of the above) (Ingold 2013a and 2013b. See also Heneise; High; and Merz, this volume). And yet cryptids arguably also instil a sense of fear in those who see them or believe in their existence, be it a fear of the unknown or a fear of what these 'unknown' creatures are capable of! They are also, in this fieldwork context at least, experienced as real, regardless of their symbolic potency.

The increasing propensity for humans to keep non-endemic or 'exotic' animals as pets in many parts of the 'developed' or 'post-domestic' world (for example Jaclin 2013), especially animals such as big cats, bears, chimpanzees or alligators who can pose a very real threat to human physical safety and well-being, could be viewed as a form of risk-taking enacted by individuals who feel particularly alienated from society. Indeed, the large body of anthropological and socio-logical material on risk-taking (e.g. Lyng 2004; Wacquant 2004), building on seminal works such as Beck (1992), Douglas and Wildavsky (1983) and Giddens (1991) suggests that in taking risks and negotiating them successfully individuals are able to regain some form of control over their otherwise chaotic lives (see also Attala, this volume). They are also able to attain status as a result of their close associations with potentially dangerous animals (see also Goldman and Walsh, this volume), as clearly evidenced in the relationships between certain disempowered youths and stigmatised 'dangerous' dog breeds which is widely documented (e.g. Burley 2008; Caglar 1997).

In my previous discussion of non-endemic big cat sightings (Hurn 2009), I argued that big cats sighted in the part of rural west Wales where I have con-ducted fieldwork over a period of twelve years, become metaphors for disen-franchised Welsh farmers, enacting a ritual of resistance which emulates the struggle for sovereignty of 'the Welsh' in the political landscape created during the ongoing process of devolution. I maintain that sightings confer a certain status on individuals within a community united by antipathy to external legis-lation designed (so it is thought) to curtail rights and destroy livelihoods. The 'origin myth' of ABCs in my fieldwork context casts them as former exotic pets who were also subjected to English legislation (the 1976 Dangerous Wild Animals Act – see also Franklin, this volume). Their mythological status hinges on the fact that sightings imply some individuals were able to overcome attempts to legislate against them by evading capture and surviving in the Welsh hills. Their ability to survive against the odds renders them iconic symbols of resis-tance and freedom. These animals have achieved what many of my Welsh farming informants continue to strive for. Having attended the 2002 Liberty and Livelihood March in central London (accompanying a coach-load of Welsh farmers), which saw rural dwellers descend on the (English) capital in their thou-sands to show support for a host of 'rural' issues and protest against the proposed ban on fox hunting, and having attended hunt functions and meets in the run-up to and following the 2004 Hunting Act, which saw fox hunting with hounds banned, I can recognise that for the farmers in my fieldwork context, ABCs can indeed be viewed as highly politicised animals, 'as a kind of return of the repressed [...] celebrated as forms of "defiance and resistance"' (Wilbert 2006: 45). The fact that official bodies refute claims and sightings makes locals all the more adamant that they are right. Therefore, it would appear that the presence of ABCs, who has seen them, and who believes their stories, has become sympto-matic of wider conflict, and sightings become yet another marker of 'belonging' in a rural community, or simply a lived reality for individuals united by a shared mode of subsistence or lifestyle (see also Franklin, this volume).

Authoritative sources and the burden of proof

Much of the public and academic debate over cryptids concerns whether or not they exist, and if their existence can be proved, determining what type of creature(s) they are (see Forth; Walsh and Goldman; and Turner, this volume). Eberhart's definition of cryptozoology as 'the study of the evidence for animals that are undescribed by science' (Eberhart 2002: xlvii) immediately prioritises 'scientific' discovery and taxonomic classification over folk sightings and experiences. This emphasis on evidence which can satisfy scientific rigour is significant. Science is, after all, widely perceived and promoted as the arbiter of truth and the need for scientific accountability certainly has had, and continues to have, implications for how individuals within this particular fieldwork context, and many of the others described in this volume, think about the presence of cryptids in their lives.

It is perhaps worth noting here that my interest in and subsequent investigation of ABC sightings was largely serendipitous. When I set out to study a small farming community in west Wales for my doctorate in 2000 I certainly did not envisage any encounters with cryptids. However, not only did my hunting and farming informants let slip periodically that they had seen big cats locally, leading me to attempt to systematically collect their oral testimonies, but forensic taphonomist Ros Coard of the University of Wales Trinity Saint David had, unbeknown to me at the start of my fieldwork, been conducting forensic analysis on the carcasses of prematurely deceased livestock found in my fieldwork area in order to ascertain whether or not they bore the hallmarks of big cat kills (Coard 2007).

I initially found myself making value judgements as to the 'authority' of each individual informant and the resultant reliability or truth of their sighting experience. If I'm honest, my concern when the subject of big cat sightings arose was that I was the victim of an elaborate ruse (see also King, this volume). However, as the number of sightings recounted to me increased, I began to take them more seriously. In many respects this sort of value judgement goes on whenever people claim to have seen or experienced things which might appear inexplicable to the uninitiated, as most of the contributions to this volume demonstrate. I also found myself the object of scepticism during an interview about my research on the BBC Radio 4 programme *Thinking Allowed*. In his closing comments the interviewer, Laurie Taylor, jocularly urged his listeners to get in touch if they too had experienced glowing eyes or things going bump in the night. In the wake of that interview being broadcast, however, I received numerous e-mails and phone calls from listeners across the UK who had indeed seen what they thought were big cats or the hallmarks of big cat predations. Many had kept their experiences to themselves precisely because of a fear of how their accounts would be received by others.

In a previous publication on big cat sightings in west Wales, and firmly under the influence of post-enlightenment ontology, I concluded that the animals my informants claimed to have seen could be explained in 'rational' terms

(Hurn 2009) – a view supported by Coard's scientific evidence. These cryptids might reasonably be real 'flesh and blood' animals. My process of rationalisation and the eventual conclusion in my own mind that ABCs were indeed real was a combination of science, folk knowledge and personal experience, including Coard's forensic evaluation; the patterns of predation which my informants described (including sheep carcasses lodged in forked tree branches several metres above the ground); the sheer volume and consistency of sighting experiences; and, most significant of all, my own encounter with a (suspected) ABC. The latter, which occurred late one night as we walked through a large field, certainly felt very real for both me and my dog Max (a large German Shepherd) – we were terrified! On our nightly walks Max habitually ran on ahead enjoying the smells and sounds of nocturnal wildlife, but on this occasion something stopped him in his tracks and he yelped and bolted back to me hackles raised, tail between his legs and trembling. I shone my torch in the direction from which he had come and caught the gleaming eyes of a large, predatory animal in the beam – predatory as the eyes were positioned so as to be looking straight at us (as opposed to on the side of the head as was the case with the 'prey' animals of that size we usually encountered – mountain ponies and sheep) – and set within a large, dark body. I had spent some time conducting participant observation with lampers (hunters who hunt at night using lamps and rifles) in the area, and was familiar with the signature reflections of the *tapetum lucidum* (the layer of reflective tissue at the back of the retina) from different animal species including foxes, badgers, rabbits, domestic cats, dogs, sheep and horses. These eyes were different. They were higher off the ground, larger, and the reflection was much yellower than any I had seen before. Max's reaction too convinced me that this was not an animal he had encountered previously (and was not one he wanted to encounter again!).

Thereafter my interviews with informants took a different turn. I found that people were much more forthcoming with detailed information if I shared our (mine and Max's) experience. Up until that point I had received lots of sightings accounts but it wasn't until I was one of the initiated that people began to alert me to the historical precedent for big cats in the area (see also Bernard, this volume). According to these narratives (or perhaps more appropriately 'foundation myths'), a local scrap dealer kept two melanistic leopards as exotic security guards at his yard. These animals mysteriously disappeared in the late 1970s or early 1980s (I received conflicting accounts regarding the date of their disappearance) in the wake of the 1976 Dangerous Wild Animals Act. As one informant who had visited the site as a child in the 1970s recalled: 'He had this huge black cat on a chain. It scared the s*** out of me. Better than a guard dog any day!'

As my fieldwork progressed and the number of sightings I had collected reached the 100 mark, I began to realise that the presence of these animals was widely accepted by my informants at an informal, community level. This was partly because if individuals hadn't seen an ABC themselves, they knew someone who had. The authorities on the other hand have, until recently, felt very differently. However, in 2011 Pembrokeshire County Council issued a statement confirming the presence of big cats in the area. The statement followed the sighting of a large black felid reported to the police and the media by a member of the council staff. What made this sighting particularly credible, according to media reports and the statement issued by the council offices, was the fact that the individual in question was a former police officer, and therefore a reliable and authoritative witness (see also Monbiot 2013). The following passage is from a statement released by the council and published in the national press, and has been quoted in full to reveal the level of detail included in a bid to convince the wider readership:

> Mr Disney's encounter happened in broad daylight in countryside six miles north of Haverfordwest, near Treffgarne village. He was driving his council car on a single track road at 15mph when a large black 'puma or panther' crossed five metres in front of him. 'I immediately stopped my vehicle and stared at this animal. It had a large cat-like head, muscular build and was approximately three feet tall,' he said. 'It was bigger and more muscular than a German Shepherd dog. The coat was smooth and looked like it had brown spots on it. 'I had a clear, unobstructed view of the animal and the visibility was excellent. The animal was in my view for fully five to six seconds, the time it took to cover the width of the road and then disappear into the undergrowth at the side.' He added: 'I am 100% certain that this was a puma or panther-like animal and was definitely not a dog, cat or any other domestic animal. It was not something I had seen before other than in a zoo.' He called in at a local farm soon after and was told that a large puma-like animal was nearby a few weeks earlier. Chief Inspector Steve Matchett of Dyfed Powys Police said: 'We are aware of a possible big cat sighting in the Treffgarne area of north Pembrokeshire, which occurred on Wednesday. We're working with all relevant agencies including Pembrokeshire County Council and the Welsh Assembly Government's Big Cat Sighting Unit has also been informed. While the public should not be alarmed by this latest possible sighting, we would urge anyone who does see what they think might be a big cat not to approach the animal and to stay a safe distance away from it.' The last publicised incident in the county was in November when a sheep carcass was reported to bear the hallmarks of a big cat kill. Pembrokeshire Council's public protection chief Mark Elliott said: 'We believe this is the closest anyone has been to a big cat in the wild and is further proof that there is at least one large animal roaming free in Pembrokeshire.'
>
> (Stone 2011: n.p.)[2]

The sighting happened in 'broad daylight', and the witness was driving an official vehicle. Exact measurements were given (such as the driving speed, and the distance between the car and the animal) all in a bid to communicate that this was, indeed, a plausible event. In a subsequent interview with *Guardian* columnist George Monbiot, Mr Disney reiterated the validity of his sighting:

If I'd been dreaming or thinking about them at the time, it might have been another matter. But it was the last thing on my mind. I was just driving along – and one crosses the road. He was probably about 3 feet high and six feet long. I would say bigger than a medium-sized dog, but definitely not a dog. He was powerful-looking, with a black, glossy, shiny coat, incredibly muscular, like a horse's shoulders.

(In Monbiot 2013: n.p.)

Mr Disney used a German Shepherd dog as a comparative for the black felid who had crossed his path, and during my own fieldwork many informants also observed the similarity between the animal they saw and a big dog, often pointing to Max, who accompanied me everywhere, to indicate approximate proportions. For example, one man described driving up the track to his farm and seeing a large, black, athletic animal loping across the field in front of him before disappearing into the adjacent woodland:

It was only a few seconds it was there. I thought it was a big dog [gestures in Max's direction], but it had a long, thin tail and it didn't move like a dog. My neighbour's got a Huntaway [another large breed of shepherding dog] so I phoned her soon as I got in the house and told her it'd escaped. She said it hadn't, it was sat with her in the kitchen! She said it must've been the 'Beast of Bont!' and we 'ad a good laugh.

I pushed him further. 'Well, I dunno what it was. It looked like a f***in' big black dog, but not quite, you know?' He went on to say that he had brought all of his 'stock' (i.e. livestock – in this case, sheep and horses) to bed down in the barn that evening, just in case.

Interestingly, black dog apparitions are common in many parts of the world and throughout history. In all contexts where sightings have been recorded of them these apparitions are typically seen as portents of imminent death (hence the belief that they are the guides who escort the recently deceased to *Annwn*, the underworld of Welsh mythology), or as harbingers of bad luck (Sherwood 2000). Certainly for many farmers a big cat sighting could lead to livestock predation which could result in death (of sheep) and bad luck (for both sheep and farmer). However, while informants frequently likened the subject of their sightings to big, black, enigmatic dogs, there was always something about the creature (e.g. the way it moved, its musculature, its tail) which told them it was not a dog.

Anthropological knowledge, scientific knowledge and the liminality of the unknown

During the course of my fieldwork I visited the local scrap yard my informants had repeatedly mentioned, hoping to interview the dealer, and maybe catch sight of something linking him to the apocryphal leopards he was rumoured to have released. However, I was greeted by several large, vociferous (almost black) dogs who hurled themselves at the chained iron gates, but no sign of the man himself (nor his feline companions). Coincidentally a few weeks later he too disappeared without trace; according to the local press he had moved to Ireland so as to avoid a court appearance. This was frustrating bad luck as it meant I was unable to verify my informants' memories! Nonetheless, the scrap yard itself represented the ideal place to have been the former abode of mysterious big cats. The piles of tyres, heaps of rusting metal and the scrap man's derelict caravan created a sense of liminality and of transience. This was not a fixed abode, and the culmination of so much matter out of place emphasised the betwixt and between nature of the scrap dealer and his mode of subsistence in relation to the neighbouring farms, whose inhabitants had deep roots and whose livelihood was the 'norm' in the area.

Attributing the presence of cryptids to a liminal or non-normative local is a popular trope in many other ethnographic contexts too. In their work on the Zanzibar leopard for example, a creature thought to be extinct as far as the scientific community is concerned, Martin Walsh and Helle Goldman (this volume) discuss the phenomenon of leopard keeping whereby witches control zoologically extinct leopards to do their evil bidding. Cryptozoological entities are also frequently associated with liminal places and spaces: bogland, waterways, crossroads and gateways are all common locations where sightings occur or where they have occurred historically. Examples include the *kelpie*, a creature widely documented in the myths of the British Isles, and its specifically Welsh equivalent, the *ceffyl dwr* (water horse). Like the fish-tailed beings of South Africa (Bernard, this volume) but taking the shape of a horse, *kelpies* or *ceffylau dwr* are typically encountered along river banks or on bogs where they lure their victims to watery graves. It is easy to rationalise the existence of *ceffylau dwr* in functionalist terms, as a cautionary tale designed to keep children away from potentially dangerous places where real flesh and blood ponies are known to dwell. Having regularly encountered the perils of Welsh bogs during fieldwork, including one unforgettable occasion when the horse I was riding sunk up to his shoulders and flailed in the mud for over half an hour before he was able to pull himself free, I can appreciate why such cautionary tales are important! However, like the landscapes in which they roam, cryptids themselves are liminal because they exist somewhere between the 'known' realm of zoologically recognised, tangible creatures (e.g Welsh ponies, leopards or dogs) and the scientifically unknown and intangible realm of ghostly apparitions and mythical creatures. Indeed, even though there is a plausible foundation myth, the number of big cat sightings in my fieldwork area is statistically significant, and Coard's forensic analysis suggested (on the basis of tooth pits in sheep and horse carcasses) that the animals she examined had been predated on by a 'medium sized felid', ABCs remain liminal because it is impossible to accurately establish key details such as their species and population size. Consequently they remain 'unknown' to science.

While the liminal nature of cryptids makes them a particularly interesting focus for anthropologists, the 'need' to classify and attempt to ascertain whether they are 'real' or whether they are merely symbolic figments of collective or individual imaginations has made them unattractive academic subjects overall. Ingold (2013a and 2013b) highlights the roots of this problem, arguing that the legitimacy of contemporary science is founded on the pursuit of 'truth' via the validation and revalidation of so-called objective data, with particular consequence for anthropologists who have,

> for the most part [...] ridden on the back of the same enterprise. That is to say, they have colluded in the division between [...] the reality of nature that can be discovered only through systematic scientific investigation, and the various imaginary worlds which people in different times and places conjured up and which in their ignorance of science and its methods they have taken for reality.
>
> (2013a: 36; 2013b: 734–735)

Ingold suggests that so long as mainstream science is concerned with 'knowing nature' as an objective reality, separated from our subjective experience, then scientific knowledge and understanding of the world will remain inherently limited. Only when we 'ground knowing in being', when our knowledge of the world emerges out of our experience of living as active subjects within the world, can we really 'know' in any meaningful sense. Although he doesn't refer to them as such, Ingold argues that encounters with cryptids (he uses the examples of medieval dragons and the Ojibwa *pinési* or Thunderbird) can heal the rupture he perceives 'between the real world and our imagination of it' (2013b: 749).

Not all anthropologists are necessarily guilty of the collusion to which Ingold refers. Certainly the recent ontological turn in anthropology has prioritised the diverse realities that exist for our equally diverse ethnographic subjects and repositioned scientific ontology as just one of many ways of seeing the world (e.g. Kohn 2015).

The suggestion that big cats roam the Welsh hills, and the accompanying sightings narratives, typically elicit derisive, sceptical responses from most outsiders including establishment science. In contemporary Wales scientific knowledge is authoritative knowledge, and consequently the tangible 'real' world of zoological species is clearly separated from the intangibility of the 'imaginary' world of apparitions and mythical creatures. Indeed, notwithstanding the case of Mr Disney, there is a reluctance on the part of authorities and the wider general public to accept that ABCs exist even in the face of mounting evidence, including the scientific kind (Coard 2007). However, for locals in this fieldwork context, ABCs are very much 'creatures of experience' and part of the real and 'natural' world, of their world, and of a different order to other cryptids who inhabit the same landscape and yet are very much creatures of the imagination (cf. Ingold 2013b: 738; 2013b: 41; and 2000: 278–279).

Nonetheless, it might be argued that ABCs occupy a middle ground between what Ingold (2013a and 2013b) has referred to as 'knowing and being' or 'science and imagination' which once again renders them liminal but also anthropologically useful beings. This middle ground is also occupied by other transgressive creatures. Indeed, non-normative behaviours enacted by scientifically recognised and verifiable animals can also turn them into mythical beings. The so-called lions of Tsavo for example, whose mythic reputation as calculating, bloodthirsty man-eaters went before them and served as justification for their subsequent extermination and exhibition as taxidermied specimens (Wilbert 2006: 34). The middle ground occupied by ABCs is also a good place to emphasise the importance of both scientific and 'folk' or experiential ways of knowing and being. Indeed, many anthropologists whose work explores the anthropology of religion have considered the beliefs of their informants to be of interest not because they are true per se, but because their informants believe them to be so. For example, according to Evans-Pritchard:

> [the anthropologist] is not concerned *qua* anthropologist, with the truth or falsity of religious thought. As I understand the matter, there is no possibility of his *knowing* whether the spiritual beings of primitive religions or of any others have any existence or not, and since that is the case he cannot take the question into consideration. The beliefs are for him sociological facts, not theological facts, and his sole concern is with their relation to each other and to other social facts. His problems are scientific, not metaphysical or ontological.
>
> (1965: 61–62)

Such an approach is also clearly visible in Kay Milton's work on environmental conservation (2000) where she suggests that anthropologists need not concern themselves with the 'reality' of their informants' beliefs, but rather should record what they say and do, and attempt to analyse and account for those beliefs. Yet it seems that whether our informants' beliefs are true or false has numerous implications for our subsequent understanding of events and so in practice we often do concern ourselves with just that, seeking to root out and present the 'truth'. This truth is based on the subjective knowledge we (anthropologists) have acquired in the field and through our academic training, in addition to our records of what our informants say and do. However, even in the wake of the reflexive and ontological turns it is sometimes difficult to avoid prioritising our own interpretations, our own realities, projecting them onto our informants, and in the process claiming an authoritative perspective.

This is something both Willerslev (2007: 19) and Ingold (2000: 76) have previously noted as problematic in relation to anthropological 'rationalisations' of animism:

When hunters use terms drawn from the domain of human interaction to describe their relations with animals, they are said to be indulging in metaphor (Bird-David, 1992). But to claim that what is literally true of relations among humans (for example, that they share), is only figuratively true of relations with animals, is to reproduce the very dichotomy between animals and society that the indigenous view purports to reject. We tell ourselves reassuringly that this view the hunters have, of sharing with animals as they would with people, however appealing it might be, does not correspond with what actually happens. For nature, we say, does not *really* share with man [and, for example, horses do not really drag humans to their watery graves]. When hunters assert the contrary it is because the image of sharing is so deeply ingrained in their thought that they can no longer tell the metaphor from the reality. But *we* can, and we insist – on these grounds – that the hunters have got it wrong.

(Ingold 2000: 76)

Before my own ABC encounter and the collation of sightings narratives and forensic data I was reasonably confident that ABCs were nothing more than creatures of the imagination – symbolically redolent metaphors for disenfranchised Welsh farmers. For that short space of time I was guilty of the sort of ethnocentric rejection of indigenous or embodied knowledge which proponents of the ontological turn have foregrounded and which Ingold critiques in the quote above. However, my resultant belief in these cryptids is not simply the result of knowing them through experience, as even though the experience was visceral, I doubted myself in the cold light of day. Rather, they have become real through the combination of my embodied experience, the shared experiences of others, the outcomes of post-mortem analysis of livestock carcasses, patterns of predation, and the recurring stories surrounding the scrap dealer and his feline companions. This combination of different methods, of different ways in which cryptids can be known, has led to ABCs becoming real as far as I am concerned. Cryptozoology is not, therefore, just the search for animals that are unknown to science. It can and should be about the process through which cryptids come to be known, and they come to be known by the variety of means by which we come to know about any other being in the world. They are, therefore, important but neglected anthropological subjects.

Conclusion: cryptid conservation?

Cryptids, along with creatures who fit comfortably into taxonomic categories, are 'known' by many people in many places 'by their traditions', i.e. the stories which people told about them, and which continue to unfold through time (Sax 2001; 2013; Ingold 2013a; 2013b). Both Ingold (2013a; 2013b) and Sax (2001) note that every creature, whether real or imagined, has a tradition and this tradition continues when its name is uttered, or if its call is heard, or if it is sighted, or if it acts. Moreover, these animal traditions are of particular significance in the Anthropocene, when humans have 'far more power than wisdom' because '[t]o preserve an animal as a tradition, we must know it intimately, we must be familiar with the lore that has grown up around the creature since time immemorial' (Sax 2001: xi).

As per the *molimo*, a creature of Mbuti mythology described by Turnbull (1961), whose voice emanates from the forest in response to human calls during ritual celebrations, some cryptids tend to have what might be termed a performative identity in the sense that their reality is confirmed when individuals encounter them (or are familiar with the encounters of family members or others whose testimonies they trust, or when science or local history provides evidence or precedent). Ingold argues that as individuals move through landscapes, interacting with other animals, plants, topographical features and weather fronts, their knowledge of the natural world is formed. Individuals who encounter ABCs may process their encounters in relation to their knowledge and experience of the natural world – comparing the animal out of place with familiar animals (e.g. dogs or horses) who they would expect to encounter in rural west Wales! The 'reality' of cryptids then is largely dependent on each individual's experiences not just of them, but of 'nature' more generally. Encounters with cryptids arise from or rather constitute what Ingold has termed 'performative knowledge'. Drawing on Ojibwe performative knowledge of another cryptid, the *pinési* or Thunderbird, Ingold (2000) observes that for the Ojibwe, thunder is the cry of *manitou* (spirits). Humans have the capacity to understand their cries as spoken language, because for the Ojibwe 'the truth of things is not only found but also tested by personal oneiric experience' (2013a: 42), and many Ojibwe have 'heard' or received messages from thunder. Ingold suggests that this in stark contrast to 'Western' knowledge of thunder which emphasises the physical mechanisms of shifting and unstable air currents, as discovered in laboratory experiments. Maybe the reality of thunder lies somewhere between the two positions? Certainly for many of the cases explored in this volume cryptids can be explained or accounted for differently according to different perspectives. However, this relativism causes problems for those concerned with classifying the world, as anthropologists often are (Ingold 2013a and 2013b).

Ingold (2013a and 2013b) suggests that there was a departure in the European intellectual tradition from being informed by nature (and human experiences in and of the natural world) to the prioritising of scientific knowledge at the onset of Modernity. In such a context there was no longer any place for dragons or other fantastical beings other than as cautionary tales and symbols of human achievement or folly. Rather than being known performatively by traditions founded on direct experience of them (or stories of past experiences of them), creatures were classified into bounded entities, i.e. species. Any who were known by tradition alone therefore, such as dragons, and whose presence or existence could not be corroborated by science, became problematic and obsolete.

The ABCs I encountered second-hand during fieldwork via their traditions, and the individual ABC Max and I encountered directly on our nocturnal walk, remain 'unknown' to science, despite Coard's forensic data, because of the lack of a captured specimen who could then be examined, DNA-tested and located within a particular taxonomic group. At present all that can be 'known' is that there is probably at least one medium-large sized felid with dark pelage, who has escaped or been released from captivity, roaming the hills and valleys of West Wales. Welsh ABCs (or at least this particular ABC) remain creatures of the imagination but not imaginary creatures (cf. Monbiot 2013). They are known both by the humans who encounter them and by the sheep and horses (and other animals) whose lives sustain theirs. However, rather than wanting to find and kill ABCs in retaliation for livestock losses as might be expected, the farmers and other rural dwellers who encounter them (in this context at least cf. Franklin, this volume), feel strongly they should be protected, precisely because of their symbolic potency – they are admired for their ability to survive against the odds. King (this volume) argues that it is important for anthropologists to protect the cryptids they encounter in the field, and certainly in this case too the conservation of ABCs is synonymous with the conservation of local ways of life, and local ways of being in the world. They have an origin myth, and a tradition within local rural communities, and their liminal reality – as unknown yet knowable zoological as opposed to imaginary beings – demonstrates that scientific and other ways of knowing are not mutually exclusive, and that cryptids (like other matter out of place) should be taken as seriously by anthropologists as they are by our informants in the field.

Notes

1 There is fossil evidence for dinosaurs in Wales (e.g. Galton et al. 2007) and there is precedent for fossilised dinosaur remains being interpreted as mythological beings such as dragons (e.g. Mayor 2001).
2 While big cat sightings were accepted within the farming community, the newspapers reporting on Mr Disney's experience also covered urban areas where big cat sightings were not typically made (see also Franklin, this volume).

References

Beck, U. 1992. *Risk society: Towards a new modernity.* London: Sage.
Burley, S. 2008. My dog's the champ: an analysis of young people in urban settings and fighting dog breeds. *Anthropology Matters* (online), 10(1), n.p.
Caglar, A.S. 1997. 'Go go dog!' and German Turks' demand for pet dogs. *Journal of Material Culture*, 2(1), 77–94.
Coard, R. 2007. Ascertaining an agent: using tooth pit data to determine the carnivore/s responsible for predation in cases of suspected big cat kills in an upland area of Britain. *Journal of Archaeological Science*, 34(10), 1677–1684.
Douglas, M. 1996. *Thought styles: Critical essays on good taste.* London: Sage.

Douglas, M. and Wildavsky, A. 1983. *Risk and culture: An essay on the selection of technological and environmental dangers.* Berkeley: University of California Press.

Eberhart, G.M. 2002. *Mysterious creatures: A guide to cryptozoology.* Santa Barbara, CA: ABC-CLIO.

Evans-Pritchard, E.E. 1965. *Theories of primitive religion.* Oxford: Clarendon Press.

Galton, P.M., Yates, A.M. and Kermack, D. 2007. Pant y draco. n. gen. for the *Codontosaurus caducus Yates*, 2003, a basal sauropodomorph dinosaur from the Upper Triassic or Lower Jurassic of South Wales, UK. *Neues Jahrbuch für Geologie und Paläontologie-Abhandlungen*, 243(1), 119–125.

Giddens, A. 1991. *Modernity and self-identity: Self and society in the late modern age.* Stanford, CA: Stanford University Press.

Hechter, M. 1977. *Internal colonialism: The Celtic fringe in British national development, 1536–1966.* Berkeley: University of California Press.

Hurn, S. 2009. Here be dragons? No, big cats! Predator symbolism in rural West Wales. *Anthropology Today*, 25(1), 6–11.

Ingold, T. 2000. *The perception of the environment: Essays on livelihood, dwelling and skill.* London: Routledge.

Ingold, T. 2013a. Walking with dragons: an anthropological excursion on the wild side. In *Animals as religious subjects: Transdisciplinary perspectives*, edited by C. Deane-Drummond, R. Artinian-Kaiser and D.L. Clough. London: Bloomsbury, 35–58.

Ingold, T. 2013b. Dreaming of dragons: on the imagination of real life. *Journal of the Royal Anthropological Institute*, 19(4), 734–752.

Jaclin, D. 2013. In the (bleary) eye of the tiger: An anthropological journey into jungle backyards. *Social Science Information*, 52(2), 257–271.

Kohn, E. 2015. Anthropology of ontologies. *Annual Review of Anthropology*, 44(1), 311–327.

Lyng, S. (ed.) 2004. *Edgework: The sociology of risk-taking.* London: Routledge.

Malinowski, B. 2004 [1948]. *Magic, science and religion and other essays.* Whitefish, MT: Kessinger Publishing.

Mayor, A. 2001. *The first fossil hunters: Paleontology in Greek and Roman times.* Princeton, NJ: Princeton University Press.

Milton, K. 2000. Ducks out of water: nature conservation as boundary maintenance. In *Natural Enemies. People–Wildlife Conflicts in Anthropological Perspective*, edited by J. Knight. Oxford: Berg, 229–246.

Monbiot, G. 2013. *Feral: Searching for enchantment on the frontiers of rewilding.* London: Penguin.

Sax, B. 2001. *The mythical zoo: An encyclopedia of animals in world myth, legend, and literature.* Santa Barbara, CA: ABC-CLIO.

Sax, B. 2013. *Imaginary animals: The monstrous, the wondrous and the human.* London: Reaktion Books.

Scheper-Hughes, N. 2001. Commodity fetishism in organs trafficking. *Body and Society*, 7(2–3), 31–62.

Sherwood, S.J. 2000. Black dog apparitions. *Journal of the American Society for Psychical Research*, 94, 151–164.

Stone, A. 2011. Big cat encounter convinces council chiefs. *Independent*, 23 October. (Accessed from www.independent.co.uk/environment/nature/big-cat-encounter-con vinces-council-chiefs-2197390.html).

Turnbull, C. 1961. *The Forest People*. New York: Simon & Schuster.

Wacquant, L. 2004. *Body and soul*. Oxford: Oxford University Press.

West, H.G. 2005. *Kupilikula: Governance and the invisible realm in Mozambique*. Chicago: University of Chicago Press.

Wilbert, C. 2006. What is doing the killing? Animal attacks, man-eaters, and shifting boundaries and flows of human–animal relations. In *Killing Animals*, edited by the Animal Studies Group. Urbana: University of Illinois Press.

Willerslev, R. 2007. *Soul hunters: Hunting, animism, and personhood among the Siberian Yukaghirs*. Berkeley: University of California Press.

12 Digesting 'cryptid' snakes

A phenomenological approach to the mythic and cosmogenetic properties of serpent hallucinations

Luci Attala

Introduction

Ancient mythological stories about extraordinary snake-beings are recounted in nearly every culture in the world (Willis 1990), which prompts the questions: What is it about snakes that make them such ubiquitous mythic tropes? Why is it that snakes consistently represent remarkableness? And, furthermore, why do phenomenal snakes with anthropomorphic abilities persist as potent signifiers in a range of cultural contexts today? With increasing numbers of people choosing to pay to experience the company of cryptid snakes in hallucinatory form, this chapter considers the contemporary significance, the tenacity, and the value of snakes through unpacking the coalesced meanings embedded in, and embodied, by snakes as they merge with human bodies. Using Damasio's (2000) ideas concerning the neurobiological purpose of emotions, 'snake' is revealed to be a sign that provokes specific physical sensations (feelings) with a view to integrate, ingest or 'stomach' the existentially troubling and socially important complex idea-visions that humans find difficult to digest.

Accounting for snakes in myths

Myths are stories that explain the unexplainable. They have been described as ideological mechanisms that out-picture cognitive structures, as metaphorical, sacred narratives, euphemistic, allegorical (Campbell 1991), truths (Eliade 1963) and as 'logical models to overcome contradiction' (Lévi-Strauss 1955: 443). Their fantastic content renders impossible feats (e.g. shape-shifting) possible, normalising the enigmatic and demonstrating solutions for existential risks by pushing on the boundaries of what is achievable. Myths open up thinking to new possibilities, suggesting there are greater alternatives to what we know, that rules can be broken and that differences exist. Mythscapes are lands of opportunity and vision, but why do so many snakes inhabit this land?

Interdisciplinary scholarship agrees that snakes make attractive candidates for mythology as they lie 'deep in the human subconscious' (Stutesman 2005: 9) from where they provoke a troubling mental mix of fascination and trepidation (Dexter 2010; LoBue and Deloache 2011). For Wagner, snakes pour themselves into the 'dark crawlspace of the unconscious' (2009: 6) where they aggravate anxieties. For both Wagner and Stutesman snakes evoke a heady mixture of fear-provoking powers (Wagner 2009; Stutesman 2005). Consequently serpentine cryptids embody perplexing and paradoxically complex assemblages of oppositional associations that circulate existential forces of creation and destruction (Dexter 2010; Stutesman 2005).

For Isbell (2009), on the other hand, any deep-rooted feelings about snakes are linked to snakes being our original predators. She claims that *ophidiophobia* (fear of snakes), which is present in all primates, has an evolutionary basis connected to our shared fruit-eating, arboreal pasts. Respect for the empirical reality of snakes, stimulated by their obvious ability to harm, could be the evolutionary precursor for the pervasiveness of snakes in myth but I wonder if this adequately accounts for the extraordinarily extensive exemplification of the serpent as a silently powerful, knowledgeable, clandestine, cosmogenetic force that is depicted in so many ancient and now modern cosmological stories (Bernard, this volume; Dannaway 2009; Lee 2007; Nicolaus 2011). Fear-of and reverence-for can, and do, accompany each other, but the phenomenological experience of snakes as a linear 'path' (Drower 1937: 37), or directional pointer, to knowing existential truths also needs explanation. It suggests that humans not only think snakes are remarkable; they *feel* they are remarkable too.

For symbologists Chevalier and Gheerbrant (1996), the material form of a snake and its particular aptitudes may hold some answers. Chevalier and Gheerbrant (1996) speak of snakes as a 'living line' – their simplicity of form allows for endless transformations and potentials. Their morphology places snakes, like the cryptozoological entities explored in the other chapters, as consistently liminal. For example, snakes are not obviously male or female, they have no legs, wings or fins, yet have been seen to fly and can inhabit virtually any environment; they stare unblinking; have forked tongues (to smell with!); are surprisingly strong, swallowing large prey whole; they can digest all types of matter without problem and they lay eggs (Stutesman 2005). In short, snakes challenge our sense of order and taxonomy and so perhaps epitomise the cognitive discomfort that Douglas believed derived from 'matter out of place' (2002). However, the anomalous pièce de résistance must surely be the snake's ability to apparently overcome death or at least prompt the illusion of rebirth by periodically shedding its skin. This process renders the snake visually haunting as the skin turns a ghostly opaque and the colour of the snake's eyes change too (Stutesman 2005). The ability to be reborn has no doubt fired humanity's imagination with most heat. It hints at a chthonic, theriomorphic aspect of immortal gods.

Incorporating snakes

Snake mythology repeats a number of key themes. Typically stories play with themes of creation but other, recurring patterns are also apparent. Of particular interest is the notion that snake and human are inextricably entangled. Characteristically, this is presented through fantastic shape-shifting abilities, but, also, through the physical incorporation of snakes into the human body or vice versa (in spiritual or material form). These representations imply that incorporation enables knowledge (snake wisdom) to be assimilated or ingested and integrated into human flesh (see Lee 2007). Through the interweaving strands of ingestion and digestion then, emerge entangled correspondences between the knower and knowing, and eating. The connection between the acts of eating, digesting and assimilating and the processes of knowing and absorbing new ideas places the snake as conduit or mechanism that joins knowledge and knowing in the body; a position that suggests the feeling of 'snake-ness' is associated with the absorption and incorporation of ideas perhaps otherwise too problematic to grasp or digest.

These themes are evident in numerous accounts. For example, members of the Hopi nation Snake Clan claim snakes as inextricably (even genetically) intertwined with humanity as the clan emerged through the union of a female snake and human male. The story asserts that, as a result of an underground search for water, a man found a community of snake people, one of whom he married. Unfortunately, the woman eventually returned to her home underground leaving their children above ground with their father. Humanity's embodied union with snake-ness in this Hopi myth reveals connections between the body and chthonic knowledge (Tyler 1964; see also Bernard, this volume).

An alternative understanding of the notion of snake as kin can be found amongst the many indigenous groups who live in southern Africa. Here various beliefs state that a snake lives within humanity as an invisible interior force named *Nyoka* (or *Nyowa*). *Nyoka's* existence is thought of as a central and controlling force within humankind. As such the human and snake are one semi-symbiotic being (Green 2004). It is widely believed that illness emerges from the discontent of the inner snake (Green 2004). According to this ontology, illness is the measure of human disrespect for materiality and is provoked when the snake recognises poor practice (Green 2004). Thus *Nyoka*, this inner snake, acts as moral compass and reminder that control is out of our hands whilst still being in our bodies. Snakes hold further significance for the South African Zulu *Isangoma* or healer/diviner. In addition to invisibly live within us, visible snakes are thought to embody the spirits of ancestors (Hambley 1931; Houle 2011; see also Bernard, this volume).

Cartesian dualities that see differences between the material and the spiritual have troubled the human mind since before the Enlightenment. The power of stories about snakes seems to come in part as an attempt to fuse these dualisms in the snake, showing its corporeality as simultaneously representative of both physical and the spiritual. Eire (2009) explains how the recurring cyclic depiction of the snake devouring itself (*ouroborus*) works with elegant simplicity to symbolise the snake's oneness by creating a singularity that describes the ungraspable sense of eternity that humans wrangle with. Eire maintains it was this that prompted Jung to suggest *ouroborus* as a central archetype for the human psyche (Eire 2009: 29). For Nicolaus (2011), on the other hand, it is the snake's elongation that suggests eternity. For other traditions this eternal cosmogenic creativity (Stutesman 2005) is represented in different ways: Shesha the Hindu cosmic serpent of eternity with 1,000 heads (Narby 1999); Mesoamerican Quetzalcoatl, the feathered serpent (Stutesman 2005); an abundance of Nordic snake goddesses, gods, witches, and world serpents as cosmic guardians (Mandt 2000); the Amazonian Barsana's ancestral anaconda that births humanity from its belly (Hugh Jones 1988; Reichel Dolmatoff 1990) and the Tukano's mythological snake person (Reichel Dolmatoff 1996) amongst others.

Another recurring theme is that of the snake as 'knower'. The snake's knowledge affords powers to alter life's course (including curatively through healing). Interestingly the theme of knowledge in association with healing is also coupled with consumption. In Christian stories for example, the snake challenges God's authority, claiming similar powers are available to all once ingestion of particular substances has occurred. There are numerous biblical references to snakes, but the story of Adam and Eve (Genesis 2) is arguably the most well known. Here the snake is pictured as villainous, using humanity's curiosity and naivety to entice them out of their paradisiacal bondage with God and into the snake's material world of personal choice. Again the human condition is reliant on a relationship with a snake, this time in association with knowledge that was originally withheld from humanity. The snake tantalises with its claim to know, promising knowledge will bring individual power. Instead, tasting knowledge proves fatal, creating the ultimate form of change – death. The snake tempts, the Bible warns, because knowledge turns out to be undesirable; it transpires that if you know, you will also suffer. Be warned, this story asserts. Don't provoke change, stay in place and change will be abated. Thus, ignorance affords permanent stability, and according to this part of the Christian story, that is paradise. The *Epic of Gilgamesh* also claims the snake stole immortality from humanity through eating. In this role reversal, the snake takes knowledge from humanity by eating a flower before the human can. It is this action that forces the burden of mortality onto the human species.

The Brothers Grimm story 'The White Snake' (1812) overturns the previous narrative to illustrate how ingesting snakes offers great power. In this story, a king's wisdom results from regular meals of snake: 'nothing … [is] … hidden from him, and it seemed as if … secret things were brought to him through the air' (Grimm and Grimm 2007: 83).

The snake's ability to moult extends the circulating themes of bodies and immortality to healing. Here we can see the shedding of body parts as suggestive of the snake's ability to cheat death, and resultantly prompts inferences of healing capabilities. To heal is also to change in repair; a reversion back to what was, rather than a destruction of what is. It is easy to see why this transformational process can be perceived as miraculous; the motivation of life to continue itself; the ability of life to know how to reform itself evokes thoughts of higher powers and life's mysteries. To fully peel off one's skin without blood or pain is a radical, fundamental alteration, comparable but also in sharp contrast to death as it replenishes and rejuvenates rather than terminates. This notion of snake as transforming healer is visible in numerous other forms. For example, the caduceus, the widely used symbol of the serpents wrapped around the staff of Hermes, signifies the sacred serpents of the priests of the cult of Asclepius, that healed injuries by licking them (Retief and Cilliers 2006).

The examples above reveal that, across cultures, the snake signifies power, creativity, realisation and transformation. Traditional stories commonly represent the snake as an all-powerful being with the ability to facilitate rebirth via the special knowledge it holds. But what of the modern interpretations associated with the idea of a snake? How do these ancient ideas translate in the minds of the modern human raised in industrialised settings and shaped by science? The following ethnographic example reveals the notion of 'snake-ness' to be as influential and as valuable a symbol now as it ever was.

Hallucinating snakes

Increasing numbers of people from industrialised nations are travelling to South America in search of the wisdom provided by cryptid snakes. These snakes promise personal healing, self-knowledge and appear as hallucinations, occurring in association with the ritual consumption of a pan-Amerindian decoction named Ayahuasca. Accounts state that communicating with the hallucinations that Ayahuasca induces provides valuable, even life-altering information (Harner 1992; Jenkins pers. comm.; McKenna and McKenna 1994; Pinchbeck 2002; Wilcox 2004).

Typing 'Ayahuasca holiday' into a search engine brings up hundreds of thousands of results. The choice of a holiday or 'retreat' that claims to spiritually recharge or cleanse is increasing in popularity (Grunwell 1998; Razam 2009). Such holidays are offered in a variety of different guises – from health spas, and hiking trips, to eco-tribal experiences. An Ayahuasca experience is one such event. The experience challenges the orthodox notion of a desirable holiday as it entails days (even weeks) of vomiting and diarrhoea, accompanied by hallucinations of snakes, sometimes of giant proportions. It is advertised as an experience for those who want to release blockages, to find their potential and to existentially heal via the therapeutic wisdom of the hallucinations.

The wisdom people seek spews forth after ingesting the revolting Ayahuasca brew in association with the guidance of a local practitioner. The main ingredient of the drink is the stem of the plant *Banisteriopsis caapi*: an endemic South American liana which twists and curls up trees, looking like a snake encapsulated in a hardened shell (Narby 1999; Razam 2009; Reichel Dolmatoff 1990; 1996; Wilcox 2004). The ritual process necessary to see the snakes is gruelling and uncomfortable. It can take weeks while seemingly endless consumption of the thick brown liquid is required (Dobkin de Rios 1972a;

1972b; 1973; 1984). The levels of active ingredients in the brew are considered toxic, which leads to purging of the liquid from both ends of the digestive system.[1] Most accounts testify that it is arduous, unpleasant and certainly not recreational, despite this activity having been labelled 'drug tourism' by Dobkin de Rios (1994). These holidays appear to be part of what is now a growing trade that recognises (and capitalises on) the wisdom and experience of indigenous ethnobotanists and their ancient methods of healing. In consequence, these 'magico-religious practices' (McKenna et al. 1998: 73) are inciting investigation into the 'brain/mind interface' (ibid.) because of the perplexing conundrum accompanying the acquisition of what is thought to be wisdom in association with hallucinations.

The ethnographic data I am primarily relying on concerns the Ashaninka of Peru, and comes from a selection of sources: the published work of Narby (1999), Razam (2009), Wilcox (2004) and Weiss (1973) combined with my own experiences and interviews with a number of informants in 2011 (Jenkins; Komaromi) and 2012 (Komaromi; Withers).

Many different Amerindian groups consume Ayahuasca for spiritual and medicinal purposes. As with other ethnographic descriptions there are discrepancies between accounts. For example, Narby's work with the Ashaninka focuses directly on the significance of the snake, suggesting that it is pivotal to consumption. But Dilwyn Jenkins (pers. comm.)[2] argues that while the snake is significant, the Ashaninka group he lived with claim its importance emerges through the neo-shamanic operations as practised in the *mestizo*.[3] This discrepancy does not detract from the fact that increasing numbers of outsiders pay to experience snake knowledge, or as Razam calls it 'snake medicine' (2009: 121) in relationship with the entheogenic properties of the brew.

The snakes experienced vary and seem tailored to individual dispositions. For example, Narby (an academic) was confronted by giant neon twin snakes that furiously hissed about his inconsequentiality (1999: 112). Komaromi, a devout and regular user, has seen friendly snakes rhythmically forming out of the air and the fire. Other accounts talk of vomit turning into rivers of snakes that pour forth from within, or as serpentine women appearing and disappearing seductively out of the matrix that manifests when everything melds together (Razam 2009: 305). My own experience was of a seemingly endless, torrential stream of dark, writhing, knotted snakes discharging through the skin of my belly – disappearing once out and away from my body. The snakes are described as both familiar and overwhelming. They can be majestic or humble as well as singular, multitudinous, terrifying or amusing. And if they talk, they share profundities that people struggle to express after the event. Significantly, the snakes are expected, even waited for, and there can be a sense of disappointment if snakes don't come (Withers, pers. comm.). It is said that you have fallen or let go when the snake arrives, succumbing to the ingredients and experience, and allowing the process to take hold (Withers, pers. comm.). When the snake arrives – in whatever form – she is commonly referred to as the *Madre Ayahuasca*: the mother of all things, and it is whilst in relationship with her that you can expect to find out more about the reality of your existence (Pinchbeck 2002: 153; Dobkin de Rios 1972a).

> I felt faced by Death, ... got nauseous, rushed out and began vomiting, all covered with snakes, like a Snake Seraph, colored serpents in aureole all around my body, I felt like a snake vomiting out the universe.
>
> (Ginsberg recounting his experience with Ayahuasca, in Burroughs and Ginsberg 2006: 60).

Desperately seeking snake wisdom?

To date only a few surveys of participants' motivation have been completed (Wilcox 2004; Winkelman 2005). From their findings it seems clear that the majority feel their hallucinatory experiences bring promise of insight into life's secrets. Winkleman (2005) and Wilcox (2004), and indeed Dobkin de Rios' earlier accounts (1972a and b, 1984), describe participants as seekers who need guidance, feel discontent, have physical or mental ailments that don't respond to allopathic treatments – in short, participants have existential issues to deal with. Whatever the motivation, the recurrent theme of wisdom as spoken or acted out by snakes echoes the conclusions of the traditional snake stories cited earlier.

However, like the biblical serpent, the Ayahuasca snakes' wisdom comes at a price, and we are warned that this journey should not be taken lightly:

> It's not a frivolous pursuit ... there's a certain amount of dread attached to taking it – you have a hallucinogenic trip that deals with death and your mortality. So it's quite an ordeal. It's not something you're going to score and a have great time on.
>
> (Sting, cited in Dunn 1998: 26)

Such statements remind us that this activity sets participants on a path or arduous passage of discovery where the snake-plant drink demands they surrender to her will. For my informants, a sense of trepidation and a feeling of risk were integral to the experience. Lyng (1990), picking up on the discourse of 'risk society' as explored by Beck (1992), Giddens (1998) and Douglas (1992), suggests that while voluntarily risk-taking appears a paradox of modern life, in fact there are significant personal rewards associated with the survival of risk. He goes further stating not only that people place 'higher value on the *experience* of risk ... [but that risk taking is] ... necessary for the well-being of some people' (Lyng 1990: 852 original emphasis).

The increasing popularity of the process suggests that the participants come away from the experience feeling that the discomfort and any accompanying risk was worth it, as these quotes illustrate:

> From my experience it's not like other drugs, not even like mushrooms, yeah? [...] When you take Ayahuasca you can see spirits right in front of

you, you're not being paranoid – they really are there. This whole experience is on a different plane than the physical. It's about healing, not escapism.

(Todd quoted in Razam 2009: 268)

My perpetual suffering is gone. I have less fear and can deal with it better. My anxiety levels are virtually gone.

(Winkelman 2005: 213)

Feeling the discomfort of the snake

So what is going on? To date conclusions have predominantly circled around the psycho-symbolic significances to account for the increasing popularity of the experience. However, investigations tend to focus on the action's efficacy, claiming value to be associated with effects. The conclusions of investigations lie between either of these two broad themes: physiological – that the extreme purging clears parasites out of the body, thus creating a sense of well-being (McKenna et al. 1998; Pinchbeck 2002; Wilcox 2004) or psychological – the hallucinations enable the individual therapeutic insight (inner sight) (Hancock 2005; Narby 1999; Pinchbeck 2002).[4] The first explanation is convincing, but slightly boring. There is evidence to suggest that purging can indeed successfully reduce parasitic populations (see Hurn 2012 on zoopharmacognosy – animal self medication and purging).[5] More problematic, however, is the fact that this explanation denies the psychological message or the symbolic communicative content of the experience and relies on the idea that humans must guard their boundaries from invaders – something that Dunn (2011) claims is the destructive obsession of many contemporary humans in industrialised nations. The second explanation is lovely but problematic. It clashes terribly with post-enlightenment scientific positivism and smacks of a hippy mishmash of ideas circling around without being able to successfully land anywhere in this jungle.

For participants, the effects of the experience do not need explanation; indeed to try and do so may detract from the potency of the realisations prompted by the communication of the visuals. Experientially or phenomenologically, the conditions are enough to confirm their 'reality', as Wilcox's description of her first meeting with Ayahuasca illustrates.

I surrendered, allowing a kaleidoscope of colour and imagery to blot out my thoughts and feeling. A huge snake appeared, dark but vibrant blue, almost indigo. It seemed much larger than I, although I could see it in its entirety. It undulated up out of the darkness, arching toward me, facing me head-on. Intricate patterns cascaded down its back in shimmering pink, green, yellow and sky blue [...] The snake just stared, alive and conscious but not wanting anything from me. There was just pure 'snakeness' hovering before me. I cannot remember exactly what happened next, but at some point the snake, at least energetically, entered my mouth and slid down my throat. I vaguely remember the odd sensation of its slithering along my spine.

(Wilcox 2004: 27)

Wilcox describes the outcome of this event as 'exquisite' because she became one with the snake or 'Mother Ayahuasca' (2004: 29). Wilcox's description is rhetorically compelling but still anticipates explanation. According to this representation: what happens, happens – not as a crazy illusion but as an ability to see reality in its genuine fullness. As one of Razam's informant's told him, Ayahuasca is teaching her to see (2009: 221).

As we have seen, the preferred interpretive direction for mythological otherness is psychological. Psychological explanations favour attributing the mechanism and value of events like these as something akin to a therapy session 'triggered by the individual's unconscious' (Razam 2009: 221). One can also assert that it is simply the extraordinary nature of the experience that is enough to account for the increasing popularity of seeking 'snake medicine' (Razam 2009: 121). But what of the regularity and homogeneity of seeing snakes? Should we not also ask what is it about visions of snakes that holds such profound communicative potency for so many different people from various walks of life?

Ashaninkan philosophy and ontology – snakes and the very substance of life

For those indigenous to the Amazon, environmental materialist explanations can account for the appropriateness of snakes and snake symbolism as the practice emerged in an environment inhabited by an abundance of numerous types of snakes. Shamanic practices work in association with animistic ontologies that envision the ecology as animated with inspirited forms. An animistic belief system is reinforced, fortified or perhaps even inspired by the visions engendered by ingesting hallucination-inducing substances. In consequence, consumers maintain that the animated hallucinations they relate with are true and real – the substance reveals a genuine vision of existence (Razam 2009). In this system, the knowledge snakes speak of exists as if within 'everything' and can be accessed on successful penetration of form's boundaries. With the help of skilled shamanic practitioners versed in the art of effective communication, perception alters to facilitate communion with the spiritual world of knowledge.

To understand this better it is necessary to grasp how these ideas work to structure Ashaninka life. In brief, Ashaninka cosmology sees all life as having once been essentially 'human'. For the Ashaninka, what now appear as distinct, different creatures are at root made out of one substance. These similarities differentiated when the great transformer, *Aveiri* (the original state), created forms to represent the variety of characters, traits, abilities and relationships that are possible (Weiss 1973). Life, through this process, was diversified into discrete, particular chunks or states, and in agreement with a number of other cosmologies, quantum or particle physics and genetic theory, all material life, or 'corporeal diversity' (Viveiros de Castro 2002: 307) is, in truth, one substance, originally undifferentiated and deriving from one place.

In this representation, form is a sign that teaches you about the things you experience (Narby 1999). In other words, morphology is both a visual cue and clue. For the shaman, awareness of shape brings accurate readings, or knowledge of who the thing really is (was), which enables solutions to be extrapolated and exposed. It is these solutions that balance health. Thus shape simultaneously hides commonality but suggests truths (Narby 1999). Here truth and knowledge are joined in a unity that comprises an essence of all things.

The Ashaninka depiction outlined above has many correspondences with other mystical stories that allow types of beings to be interchangeable. Viveiros de Castro says that 'an original state of undifferentiation' is a virtually 'universal Amerindian notion' (2002: 309). He goes on to say that this conceptual perspective inevitably 'reshuffle[s] ... our conceptual schemes' (2002: 307) because assuming a one-ness abandons dualisms and our 'dichotomous heritage' (2002: 308), declaring through 'cosmological transformation' (2002: 308) that the commonality between all things is termed 'humanness', something which carries with it a sense of personhood and agency that can be reached by utilising particular practices. Thus, it follows, 'the common point of reference for all beings ... is not humans as a species but rather humanity as a condition' (Descola 1986: 120, cited in Viveiros de Castro 2002: 309).

Challenges to taxonomic boundaries are common. Tukano beliefs, for example, are a case in point. Neighbours to the Ashaninka and also regular ritualistic consumers of Ayahuasca, the Tukano speak of a 'Snake Person' (Reichel Dolmatoff 1996: 173) who is simultaneously the mother of fish, as well as also being the river (Reichel Dolmatoff 1996: 106, cf. Bernard, this volume). This is a clear example of alternative categorisation that does not trouble over taxonomic distinctions in the same way that other cognitive and epistemological traditions do. When boundaries break down, objects blur and when this happens what might be called 'a snake' through one lens, is also simultaneously other things. Ayahuasca consumption breaks down boundaries as the plant facilitates hallucinations that not only take the form of a snake, but are also animated by the 'person' snake mother, who in accordance with this relational ontology of oneness, is, in essence 'human'.

Such ontologies confront positivistic notions of 'reality' and are commonly rejected as fantastic or allegorical, understood only as naïve representations. Kohn, however contests this, calling for the recognition of 'transspecies intersubjectivity ... [so that the] ... semipermeable membranes that constitute the borders along shifting ontological frontiers' (2007: 7) may be blurred. After living with the Amazonian Runa (a group who also use Ayahuasca), Kohn disputes positioning humans as the legitimate knowers. For Kohn, this amounts to 'cosmological autism' (2007: 9) – a state that fails to recognise the agency or subjectivities of others due to the 'physical discontinuities that separate different kinds of beings' (2007: 7). He goes further, citing Runa belief, to assert that hallucinogens may be useful to break trans-species communicative boundaries:

> Lower [beings] ... can only see the world from the perspective of higher beings via privileged vehicles of communication, such as hallucinogens, which can permit contact among souls of beings inhabiting different onto-logical realms. Without special vehicles of communication, such as halluci-nogens, lower beings understand higher ones only through metaphor – that is, through an idiom that establishes connections at the same time that it differentiates.
>
> (2007: 17)

Kohn argues that we need to allow these other thought forms to inhabit our mental worlds as traditional Amerindian groups do (2007; and see Kohn 2013 for a wider discussion on ontology and translation).

As has been shown, once the serpentine hallucinations created by Ayahuasca are seen, they are not thought of as illusions; they are accepted as real creatures now made visible, creatures who are communicators of truth and holders of knowledge. When knowledge (and truth) is assumed to be communicated through hallucinations, the simple use of a relativist framework, may not be enough (see Sperber 2000). The definition of hallucination describes imagery perceived without form as if external to the viewer. Rationalistic notions of innate knowledge might be able to shelter hallucinations under their wing through neuroscientitic notions of abnormal stimulation of the visual cortex, but how do we translate innate knowledge into apparently external, talking, ephemeral beings who recur as serpents of all sizes, colours and moods?

Epistemologically, knowledge contained within or throughout substance is perhaps even harder to conceive of. Narby (1999) wrestled with these problems, paying particular attention to snakes as hallucinations. He chose to take the experience and conclusions he had as factual with the aim of arriving at a direct understanding rather than one that might get lost in translation. In con-sequence he linked Ashaninka shamanic ideas about form with the forms that he hallucinated whilst influenced by Ayahuasca. After much deliberation he con-cluded that the shape of the snake is presented as a recurrent theme, not only in these rituals but also in a wide number of cosmological examples across the world, because it is universally significant to all life. As he sees it the snakes reflect knowledge hidden within all life forms, because life's code or message assumes the shape of 'intelligent' inter-twined snakes – our DNA (Narby 1999).

Emotionality, discomfort and snakes

Narby's (1999) claim that snake hallucinations are projections of the micro-scopic inner 'snakes' that comprise the DNA that motors every living cell is ambitious. It grounds snake hallucinations in flesh, as opposed to illusion. Using his ideas as a springboard, I would like to suggest an alternative, which links the mentality of representation to the empirical corporeality of snake by using the physiological chemistry of emotions as the communicative medium that makes sense of the impression of 'snake'.

As part of humanity's communication toolbox, cosmologies and myths are used as vehicles to convey particularly dense meanings through the imagery of the fantastic. A story without fantastic imagery can provoke emotional responses, but those with the fantastic manage to stimulate feelings (or a sense) of those grand ontological concepts that remain elusive to successful expression in other methods of discourse. Such stories allow the ideas to settle into the body, where we make *sense* of the inexpressible by having physical sensations about them – awe felt as uneasiness in the stomach, fear that makes us shrink away, happiness that comes as if bursting from the chest. Wonder-filled stories elicit emotional reactions viscerally felt, understood perhaps as a three-dimensional phenomenological picture.

According to Navarro (2008) there are two basic states of being: comfort and discomfort. All responses to life lie on this continuum. We are either comfortable or uncomfortable, or perhaps most commonly somewhere in between, and in a complex chicken-and-egg cycle of stimulus/response our equally complex and not fully understood biochemistry alters homeostatically to rebalance our physicality. Most of this is achieved in the pertinently named 'reptilian' part of our brain (the limbic system) where the amygdala helps to translate basic emotional responses into actions for survival (Maclean 1990; Pert 1997).

Emotions are simultaneously felt and communicated (Damasio 2000). Ekman's work shows that physical sensations (feelings) are directly linked to specific neurochemical pathways activated when a person explains they feel one of a number of states we have named emotions (Ekman 2004).[6] Accordingly these involuntary corporeal 'movements' mean we 'feel' emotions, or emotions' form, before we know we are feeling them (Damasio 2000; Ekman 2004). In other words, physiological chemistry is active without individual awareness, suggesting human emotional responses are involuntarily and instinctive. Darwin too, in his study on the expression of emotions (1872), unfashionably for the time, linked emotions to physiology. Rather than some sort of disassociated Cartesian idea of emotions as thoughts being located in the mind, Darwin (1872), and more recently Ekman (2004), Damasio (2000) and Pert (1997), claim that emotions are physiological processes that get translated into thought only after they begin to be felt. Pert goes further to state that the body *is* the unconscious mind because, as she sees it, the chemical communication that translates as emotion arises for distribution throughout the physiological nervous system (1997: 141) and does not appear to originate elsewhere. Accordingly, emotionality is not only initially uncontrollable, it is also stimulated by a sensorial cycle that humanity has not yet found a beginning to (Damasio 2000).

Sensations create an empirical awareness of being alive. Milton uses Damasio's work on emotion and feelings to assert that emotions are 'observable bodily changes' and feeling is 'the subjective experience of those changes' (Milton 2002: 80). This representation proposes *emotions are felt* which need not be contested except to note that while emotions are always felt and culturally judged within the semiotic code they inhabit, feelings may not always be emotional. The difference between feelings and emotions lies in the 'motivation' of the sensations (bubbling in the stomach from hunger, heat on the face from the fire, itch in the foot, etc.), these sensational signals are experienced in relationship with environmental conditions – and *might* prompt emotional feelings, but not necessarily. Thus, emotions are sensationally experienced thoughts, while feelings are sensational interactions.

Mythopoetics

Myth works to exploit emotionality too (Lewis 1955), apparently tapping into the most ancient part of the brain in a similar way to intense unusual experience, which the 'reptile' in us appears to viscerally, or perhaps without thought (unless thought is chemistry), understand (Pert 1997). Jung explored this internal, non-conscious comprehension in his work with mythology, the psyche and his theory of archetype (1981, 1997). Accordingly, Jung maintained that the psyche, as with physiology, is similarly occupied with homeostatic harmonisation; while physiology regulates bio-physical processes like tempera-ture, the psyche rebalances or regulates meaning so as to maintain a mental equilibrium. Thus, through dreams and myth, a system of classifications craft ideas into socially meaningful entities and philosophies that equalise any destabilising thoughts (Jung 1981).

For Campbell (1991) too, myth's significance lies in creating balance. Campbell rejects some of Jung's purely rationalistic stance and places myth as the stability between the inner and outer bodily experiences that being alive demands. Believing that the significance of mythological (or cryptozoological) creatures (and the snake in particular) is connected to the temporality of physical existence, Campbell claims that myths function as mirrors put in place to reflect back who you are in preparation for the two revolving, cohering eternities of new life and new death. New life looks at new death and vice versa, in a cycle of existence. According to Campbell, the serpent not only represents the dynamics of existence but also the bondage of life to time.

For Watts, however, this was not what urges us to mythologise (1989). He maintains that myth attempts to share those indescribable feelings that are, as result of embodiment, experienced alone.

> The sensation of 'I' as a lonely and isolated center of being is so powerful and commonsensical, and so fundamental to our modes of speech and thought, to our laws and social institutions, that we cannot experience selfhood except as something superficial in the scheme of the universe.
>
> (Watts 1989: 12)

Watts sees the seemingly inevitable conception of self as distinct from other parts of existence as contributing to the construction of mythology. For Watts, being 'isolated "egos" inside bags of skin' (Watts 1989: 9) provokes a tendency to leave aside any '*common* sense' (1989: 9, original emphasis). For Watts, we urge ourselves towards the sense of joining through the emotional responses provoked by myth. In conjunction with this, and by way of illustrating how words construct, he adds that it is a mistake to assert that we come *into* this world, because, in agreement with Ingold (2011) he states 'we come out of it' (Watts 1989: 9).

For Ingold, any notion that humanity exists in separation from the material world is an illusion and a dilemma that needs challenging (2011). In his earlier work, Ingold maintained that myths enable a 'poetic involvement' (2000: 56) with existence that allows the world to 'open out' (2000: 18) and bring other-than-human persons into human lives. This process, far from being a cerebral language based on 'intellectual impulse[s]' à la Lévi-Strauss (1963: 210), is one that works to draw together seemingly disparate physical threads into a meshwork that realises existential material connectivity and similarity, to dissolve the discomfort of separation and difference. Myths allow us to appreciate how the complexities of our shared worlds are interwoven with our more-than-human companions and as such provoke us to come to our senses.

Jung, on the other hand, uses the term archetype to refer to a collective, instinctual memory or inheritance that we all receive as a constituent of corporeality. An archetype is a mode of communication, or a symbolic structure that exists within the psyche as a unit of comprehension and meaning. It is with these templates of meaning that we grasp deep composite complexities, all of which allow us to negotiate the puzzle of existence. Contrary to environmental or materialist ideas, for Jung the meaning of archetypes is not derived from the 'outside'. He maintains that what is 'inside' has 'an *a priori* structure of its own that antedates all conscious experience … [because] … The psyche is part of the inmost mystery of life, and it has its own peculiar structure' (1959: 187). His claim directly connects the notion of archetypes with an existential fundamentality that has a non-material reality. This psychic reality, for Jung, is where the shapes of meaning are made, or where form, as an idea, materialises. Thus, as spheres of meaning, archetypes become motivating patterns of imagination from which the organisms or beings with supernatural qualities and abilities in myth can be found. For Jung the creatures and their powers should not be understood as illusions, disassociated imaginings or even interpretations of matter (1959: 187). He suggests that what we call illusions 'may be for the psyche an extremely important life-factor, something as indispensable as oxygen for the body – a psychic actuality of overwhelming significance' (1931: 111). Rather then, for Jung, mythologems, primordial images or archetypes organise not through interpretation or translation of the material world but instead through the out-picturing of shared hard-wired, mental patterns that exist within.

Both Campbell's life dynamics and Jungian archetypes contribute to explaining the intensity and concentration of the sign 'snake' in human culture. The brain's pattern recognition software is an equally convincing position to adopt. However, more convincing might be to link recent neuroscientific findings concerning the invisible chemical pathway of emotions with the human inability to verbally express the visceral nature of feeling and emotionality.

If the snake makes us uncomfortable, it may be used to convey other ideas that are experienced as discomfort, or that are felt to be viscerally indigestible. Stories about fantastic, cryptozoological snakes provoke this kind of gut-moving discomfort. Drinking *Ayahuasca* does the same.

Conclusion

Humans commonly use the idea of snakes to simultaneously epitomise the sinister fear of death and the awesome power of life. The duality of meanings associated with the symbol 'snake' allows snake-ness to easily embody a visceral rather than intellectual sense of eternity along with other emotionally charged existential themes that are difficult to articulate verbally. The multivalent properties associated with the emotionality of the snake increase its value as signifier, making it a key theme in cosmological stories across the world. The ambiguity embedded in mythological snakes is connected to the empirical reality of snakes in association with the attributed characteristics that the snake's form facilitates. The snake's morphological simplicity, along with its seeming ability to cheat death, advertises its potential as supernatural creature with special powers and knowledge,.

Mythological stories may not play with fantasies or the fantastic but simply re-present the world using the alternative categorisation of feeling, as if depicting the world as it is felt, rather than as it is thought. The snake's mythological potency is still active in today's post-modern world. The consumption of Ayahuasca, in the hope of corresponding with the serpentine mother, is testament to the ongoing power of the sign.

Notes

1 For more information on the ritual, see: Narby (1999); Pinchbeck (2002); Razam (2009); Wilcox (2004). Apparently, regular consumption of Ayahuasca may acclimatise individuals to the substance and stops the nausea and the snakes. (Komaromi, personal communication, September 2011).

2 Dilwyn was interviewed for this research in 2011. More information about his work with the Ashaninka can be found here: www.ecotribal.com

3 The *mestizo* is the term used to describe the border settlements made between the jungle and the town. It is here where perhaps traditional ideas have been altered and augmented to create an alternative method for the Western traveller who is in search of some link with traditional wisdom.

4 Ayahuasca is sometimes called *la purga* (Wilcox 2004: 106) in acknowledgement of its ability to cleanse the body.

5 One can see how violent expulsion of stomach and colon contents could dislodge any dwellers, and indeed, investigation shows that substances such as Ayahuasca are anthelmintic. First because they contain harmine, which exhibits significant affects on parasitic populations (Hopp et al. 1976), and second because recent studies (Max et al. 2003) also indicate that substances high in tannin reduce the parasitic burden in farm animals' digestive systems through not only increased intestinal motility but also by killing the parasites themselves. All red wood plants are high in tannin; Ayahuasca is a red wood plant.

6 It is important to note that all peoples feel and show these emotions in the same way – the physical experience and communication is identical. However, different cultures will feel angry or afraid, for instance, about different things. Thus there is some cultural difference as to what disgusts or is considered funny, but whatever prompts this, it will always be felt and shown facially in the same way (Ekman 2004).

References

Beck, U. [1986] 1992. *Risk Society: Towards a New Modernity*. London: Sage.

Burroughs, W. and Ginsberg, A. [1963] 2006. *The Yage Letters Redux* (4th edn). San Francisco: City Lights Books.

Campbell, J. 1991. *The Power of Myth*. New York: Anchor Books.

Chevalier, J. and Gheerbrant, A. 1996. *The Penguin Dictionary of Symbols*, translated by J. Buchanan-Brown. London: Penguin Books.

Damasio, A. 2000. *The Feeling of What Happens: Body, Emotion and the Making of Consciousness*. London: Vintage Books.

Dannaway, F.R. 2009. *Be Ye Wise as Serpents: A Short Meditation on Ophidian Botany* (Accessed 22 March 2015 from www.scribd.com/doc/16389117/Be-Ye-Wi se-as-Serpents-A-Short-Meditation-on-Ophidian-Botany).

Descola, P. 1986. *La Nature Domestique: Symbolism et Praxis dans L'ecologie des Achuar*. Paris: Maison des Sciences de l'Homme.

Dexter, M. 2010. The Ferocious and the Erotic: 'Beautiful' Medusa and the Neolithic Bird and Snake. *Journal of Feminist Studies in Religion*, 26(1), 25–41.

Dobkin de Rios, M. 1972a. *Visionary Vine: Hallucinogenic Healing in the Peruvian Amazon*. San Francisco: Chandler.

Dobkin de Rios, M. 1972b. Visionary Vine: Psychedelic Healing in the Peruvian Amazon. *International Journal of Social Psychiatry*, 17, 256–269.

Dobkin de Rios, M. 1973. Curing with Ayahuasca in an Urban Slum. In *Hallucinogens and Shamanism*, edited by M. Harper. London: Oxford University Press, 67–85.

Dobkin de Rios, M. 1984. *Hallucinogens: Cross Cultural Perspectives*. Australia: Unity Press.

Dobkin de Rios, M. 1994. Drug Tourism in the Amazon: Why Westerners are Desperate to Find the Vanishing Primitive. *Omni*, 16(6), 16–19.

Douglas, M. 2002. *Purity and Danger: An Analysis of Concepts of Pollution and Taboo*. Oxford: Routledge.

Douglas, M. 1992. *Risk and Blame: Essay in Cultural Theory*. London: Routledge.

Drower, E. S. 1937. *The Mandaeans of Iraq and Iran: Their Cults, Customs, Magic, Legends, and Folklore*. Oxford: Clarendon Press.

Dunn, J. 1998. Sting. *Rolling Stone*, 5 February, 26.

Dunn, R. 2011. *The Wild Life of Our Bodies: Predators, Parasites and Partners that Shape Who We Are Today*. New York: HarperCollins.

Eire, C.M. 2009. *A Very Brief History of Eternity*. Princeton, NJ: Princeton University Press.

Ekman, P. 2004. *Emotions Revealed: Understanding Faces and Feelings*. London: Orion House.

Eliade, M. 1963. *Myth and Reality*. San Francisco: Harper and Row.

Giddens, A. 1998. *The Third Way: The Renewal of Social Democracy*. London: Polity Press.

Green, E.C. 2004. Purity, Pollution and the Invisible Snake in South Africa. *Share The World's Resources* (Accessed 22 March 2015 from www.stwr.org/index2.php?option=com_content&do_pdf=1&id=113).

Grimm, J. and Grimm, W. [1812] 2007. *The Complete Fairy Tales*, translated by J. Zipes. London: Vintage Books.

Grunwell, J.N. 1998. Ayahuasca Tourism in South America. *Multidisciplinary Association for Psychedelic Studies*, 8(3), 59–62.

Hambley, W.D. 1931. *Serpent Worship in Africa*. Chicago: University of Illinois Press.

Hancock, G. 2005. *Supernatural: Meetings With the Ancient Teachers of Mankind*. London: Arrow Books.

Harner, M. 1992. *The Way of the Shaman*. San Francisco: Harper.

Hopp, K.H., Cunningham, L.V., Bromel, M.C., Schermeister, L.J. and Kahlil, S.K.W. 1976. In Vitro Anti-trypanosomal Activity of Certain Alkaloids Against *Trypanosoma Lewisi. Lloydia*, 39, 375–377.

Houle, R.J. 2011. *Making African Christianity: Africans Reimagining Their Faith in Colonial South Africa*. Lanham, MD: Lehigh University Press.

Hugh-Jones, S. 1988. The Gun and the Bow: Myths of White Men and Indians. *L'Homme*, 28(106), 138–155.

Hurn, S. 2012. *Humans and Other Animals: Cross-Cultural Perspectives on Human–Animal Interactions*. London: Pluto Press.

Ingold, T. 2000. *The Perception of the Environment: Essays on Livelihood, Dwelling and Skill*. London: Routledge.

Ingold, T. 2011. *Being Alive: Essays on Movement, Knowledge and Description*. Abingdon: Routledge.

Isbell, L. 2009. *The Fruit, the Tree and the Serpent: Why We See so Well*. Cambridge, MA: Harvard University Press.

Jung, C.G. 1931. *The Aims of Psychotherapy*. In *Collected Works Vol. 1*. Princeton, NJ: Princeton University Press.

Jung, C.G. 1959. *Psychological Aspects of the Mother Archetype*. In *Collected Works, Part 1: The Archetypes and the Collective Unconscious*. Princeton, NJ: Princeton University Press.

Jung, C.G. 1981. *The Archetypes and the Collective Unconscious: Collected Works* (2nd edn) vol . 9:1. Princeton, NJ: Bollingen.

Jung, C.G. 1997. *Man and His Symbols*. Canada: Random House.

Kohn, E. 2007. How Dogs Dream: Amazonian Natures and the Politics of Transspecies Engagement. *American Ethnologist*, 34(1), 3–24.

Kohn, E. 2013. *How Do Forests Think? Towards an Anthropology Beyond the Human*. Berkeley: University of California Press.

Lee, C. 2007. The Legend of the White Snake: A Personal Amplification. *Psychological Perspectives*, 50(2), 235–253.

Lévi-Strauss, C. 1955. The Structural Study of Myth. *Journal of American Folklore*, 68(278), 428–444.

Lévi-Strauss, C. 1963. *Structural Anthropology*. New York: Basic Books.

Lewis, C.S. 1955. *Surprised by Joy*. New York: Harcourt Brace Jovanovich.

LoBue, V. and Deloache, S. 2011. What's so Special About Slithering Serpents? Children and Adults Rapidly Detect Snakes Based on Their Simple Features. *Visual Cognition*, 19(1), 129–143.

Lyng, S. 1990. *Edgework: The Sociology of Risk Taking*. New York: Routledge.

Maclean, P. 1990. *The Triune Brain in Evolution: Role in Paleocerebral Functions*. New York: Plenum Press/Springer.

Mandt, G. 2000. Fragments of Ancient Beliefs: The Snake as a Multivocal Symbol in Nordic Mythology. *Revision*, 23(1), 17.

Max, R.A., Wakelin, D., Buttery, P.J., Kimambo, A.E., Kassuku, A.A. and Mtenga, L.A. 2003. Potential of Controlling Intestinal Parasitic Infections in Small Ruminants (Sheep and Goats) With Extracts of Plants High in Tannins (Accessed 7 November 2015 from www.fao.org/docs/eims/upload/agrotech/1905/Merida7final.pdf).

McKenna, D.J., Callaway, J.C. and Grob, C.S. 1998. The Scientific Investigation of Ayahuasca: a Review of Past and Current Research. *Heffter Review of Psychedelic Research*, 1, 65–76.

McKenna, T. and McKenna, D. 1994. *The Invisible Landscape: Mind, Hallucinogens, and the I Ching*. San Francisco: Harper.

Milton, K. 2002. *Loving Nature: Towards an Ecology of Emotion*. London: Routledge.

Narby, J. 1999. *The Cosmic Serpent: DNA and the Origins of Knowledge*. London: Orion Books.

Navarro, J. 2008. *What Every Body Is Saying*. New York: HarperCollins.

Nicolaus, P. 2011. The Serpent Symbolism in the Yezidi Religious Tradition and the Snake in Yerevan. *Iran and the Caucasus*, 15(1/2), 49–72.

Pert, C.B. 1997. *Molecules of Emotion: Why You Feel the Way You Feel*. Sydney: Simon and Schuster.

Pinchbeck, D. 2002. *Breaking Open the Head: A Psychedelic Journey into the Heart of Contemporary Shamanism*. New York: Broadway Books.

Razam, A. 2009. *Aya: A Shamanic Odyssey*. n.p.: Icaro Publishing.

Reichel Dolmatoff, G. 1990. The Cultural Context of an Aboriginal Hallucinogen: *Banisteriopsis Caapi*. In *Flesh of the Gods: The Ritual Use of Hallucinogens*, edited by P. Furst. Long Grove, IL: Waveland Press.

Reichel Dolmatoff, G. 1996. *The Forest Within: The World-View of the Tukano Amazonian Indians*. Totnes, Devon: Themis Books.

Retief, F.P. and Cilliers, H. 2006. Snake and Staff Symbolism and Healing. *Acta Theologica* (Supplementum 2) 26(2), 189–199.

Sperber, D. 2000. Metarepresentations in an Evolutionary Perspective. In *Metarepresentations: A Multidisciplinary Perspective*, edited by D. Sperber. Oxford: Oxford University Press.

Stutesman, D. 2005. *Snake*. London: Reaktion Books.

Tyler, H.A. 1964. *Pueblo Gods and Myths*. Norman: University of Oklahoma Press.

Viveiros de Castro, E. 2002. Cosmological Deixis and Amerindian Perspectivism. In *A Reader in the Anthropology of Religion*, edited by M. Lambek. Oxford: Blackwell.

Wagner, S. 2009. The Snake in the Crawlspace. *The North America Review*, 294(5), 5–8.

Watts, A. 1989. *The Book on the Taboo Against Knowing Who You Are*. New York: Vintage.

Weiss, G. 1973. Shamanism and Priesthood in Light of the Campa Ayahuasca Ceremony. In *Hallucinogens and Shamanism*, edited by M. Harner. New York: Oxford University Press.

Wilcox, J.P. 2004. *Ayahuasca: The Visionary and Healing Powers of the Vine of the Soul*. Rochester, VT: Park Street Press.

Willis, R. 1990. The Meaning of the Snake. In *Signifying Animals: Human Meaning in the Natural World*, edited by Roy Willis. London: Routledge.

Winkelman, M. 2005. Drug Tourism or Spiritual Healing? Ayahuasca Seekers in Amazonia. *Journal of Psychoactive Drugs*, 37(2), 209–218.

Index

Page numbers in *italic* refer to tables; page numbers in **bold** refer to figures.